Construction Delay Claims

Special Tutorial Edition

with

Assessments of Project Completion, Productivity, Overheads, and Profit

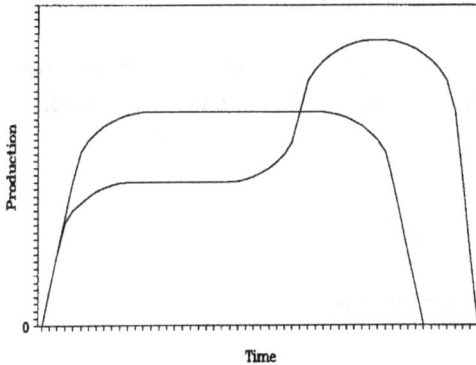

Arthur O.R. Thormann

Specfab Industries Ltd.
Edmonton, Alberta, Canada

National Library of Canada Canadian Cataloguing in Publication

Thormann, Arthur O. R. (Arthur Otto Rudolf), 1934-
        Construction delay claims : with assessments of project completion, productivity, overheads, and profit / Arthur O.R. Thormann. -- Special tutorial ed.

Includes bibliographical references and index.

bound ISBN-13   978-0-9685198-1-3
       ISBN-10   0-9685198-1-4

pbk   ISBN-13   978-0-9685198-2-0
       ISBN-10   0-9685198-2-2

        1. Construction contracts–Canada. 2. Construction industry–Law and legislation–Canada. 3. Liability (Law)–Canada. I. Title.

KE933.T48 2003   343.71 '078624   C2003-900250-0
KF902.T482003

Copyright © 2001 Arthur O.R. Thormann

Publisher: Specfab Industries Ltd.
                  13559 - 123A Avenue
                  Edmonton, Alberta, Canada
                  T5L 2Z1
                  Telephone: 780-454-6396

Cover Design: Arthur O.R. Thormann

Publication assistance by

PAGEMASTER
PUBLISHING
PageMaster.ca

Motto

To reduce claims, prepare for them!

Dedication

To Estimators,
who must anticipate the probabilities;
To Project Managers,
who often complete projects against the odds;
To Construction Owners,
who will always gain by aiding the process;
and
To the Legal Profession,
who tries to make sense of it all.

Appreciation

I also wish to extend a special thanks to
Robert Blakely, Robert Riddle, Barbara Every,
David Tettensor, Lawrence Donnelly,
Nancy Thormann, Casey Skakun,
and William Kenny
for their valuable critiques.

Preface
by
Robert S. Riddle, L.L.B.

The construction process is a complex undertaking. This is true no matter if the project to be constructed is a home, a commercial building, an institutional building such as a hospital, or an industrial facility such as an oil refinery or a hydroelectric dam. The successful completion of the project requires the contractor to plan and execute the integration of materials, labor, and services into the cohesive whole, the completed project.

As is quite often the case, when the contractor has provided a firm price for the completed project, he is obliged to work and plan efficiently in order to realize a profit for his efforts. Problems with either the planning or the execution have an impact on the final costs to construct, and an adverse effect on the expected profit. If the problems encountered result from the actions or inaction of others, then the contractor may have a claim for damages suffered due to the disruptions and delays. Conversely, the owner may have a claim against the contractor if the contractor caused a delay to the completion of the facility.

The preparation and presentation of a construction delay claim can be a daunting prospect. It can entail a level of complexity similar to the construction project itself. There are a number of experts or consultants that specialize in preparation and presentation of delay claims. However, understanding the process, the cause of the delay, the effect on profits, and how to properly assess and present a claim is useful information for a wide-ranging group of individuals including the contractor and his staff, owners, developers, design teams, quality control and assurance groups, lawyers, accountants, and insurers.

Arthur Thormann has been actively involved in the construction industry for close to fifty years. In this period, he has been an apprentice, a tradesman, a project manager of construction projects, a manager of large construction corporations, and a consultant to various groups within the construction industry. For the better part of this period, he has also been called upon to assemble, present, or resolve construction disputes and claims, including those caused by delay, or including delay components. As a result of his experiences and to assist others who may encounter similar situations, Mr. Thormann has consolidated and distilled his knowledge into the book you are about to read. He has prepared a cohesive guide designed to assist anyone who is faced with the need either to assemble or to understand the foundation of a delay claim.

Mr. Thormann begins with the estimating process and then takes the reader through the construction process: how to keep documentary records; how to identify, quantify, and address disruptions and delays; the importance of mitigating their effects; and how to recover their inevitable damages — in short, the A to Z of construction delay claims.

As a barrister and solicitor, and as a member of the Alberta Bar, my practice is primarily restricted to the areas of corporate commercial matters and construction law. Prior to entering the practice of law, I was myself actively involved in the Alberta construction industry as a contractor for many years. In the past few years, I have had the opportunity of working with Mr. Thormann on projects with delay components and have personally had the benefit of his insights and comments on the issue of delay and delay claims. Mr. Thormann has taken the time and effort to assemble his views and experience on delay and delay claims, and has thereby made his expertise available to a wider group. I heartily recommend the book to you.

Introduction
by
Author

Construction is often performed with delays and/or disruptions. When this happens, the contractor not only loses time, but also incurs additional overheads and production costs. Furthermore, because most projects are intended to produce revenue, the owner, too, incurs losses if the completion date is delayed.

To some extent, the effects of disruptions and delays can be reduced through anticipation, organization, and mitigation. However, sometimes the damages are serious enough that neither the contractor nor the owner is in a position to absorb them, and delay claims may be unavoidable.

When delay claims are unavoidable, a host of questions must be answered: What were the exact causes? Who is responsible? What were the effects? How were the damages mitigated? Were the legal requirements of the contract followed? Are the damages properly quantified? and so on. Furthermore, because the cost of disruptions and delays never produces anything tangible, nobody is happy and emotions run high.

Nevertheless, a good understanding of the issues goes a long way to help alleviate and resolve the problems. This book attempts to provide the basics for this understanding. The main points discussed in the book are as follows:

- The onus is on estimators to anticipate likely disruptions and delays and to make adjustments, both to their original tenders and to the quotations for contract changes — especially if these contract changes impact the work in progress, including that for previous contract changes.
- Every tender should include an elementary construction schedule.
- The format of the construction schedule should be designed to include expected disruptions and delays during construction and, preferably, their effects.
- It is advisable to use a construction progress measuring system in addition to the construction schedule, to more accurately measure the construction progress and any productivity losses.
- Weekly progress reports should be issued to record the work done in the past week, the work planned for the next week, and any constraints encountered. These reports can also act as

notices of the constraints.

- Productivity-loss assessments should be made as soon as the loss occurs and should be included in the delay and disruption notices.
- Project management systems should include sound practices to keep track of and evaluate disruptions and delays. These practices often help to reduce claims.
- If the damages of disruptions and delays cannot be sufficiently mitigated or absorbed, negotiations for reimbursements should begin when the losses occur and not at some later date.
- If a claim submission is unavoidable, it should be brief, to the point, and with convincing explanations of the causes and the damages incurred.
- Damages for disruptions and delays are normally limited to productivity, overhead, and profit losses, and the parties involved in a claim should be well versed in these concepts.
- When looking for the help of a lawyer or a claims expert, it pays to get the best and to make sure he or she is well informed.
- Negotiations for settlement should continue even after the claim is submitted, and one should never lose sight of the business relations angle. It is a long road that does not have a curve in it somewhere.

The subject of production losses is somewhat lengthy, and, rather than expanding the main text, I have provided more detail in an appendix. For the convenience of the reader, I have also included a glossary of construction terms and related concepts, a bibliography of related subjects, and an index.

With respect to gender, number, and generic terms, for the sake of simplicity, I have used masculine pronouns to include the feminine. This use is neither an oversight nor intended as a slight to women, whose roles in the construction industry are of ever-increasing importance. Wherever the masculine and/or singular is used and the context permits, the same shall apply to the feminine and/or the plural, and vice versa. The term *contractor* is intended, in most cases, to include subcontractors; when distinctions are necessary, the terms *general contractor* and *subcontractor* are used.

Table
of
Contents

ILLUSTRATIONS

ILLUSTRATIONS

Figures

1

The Onus on Estimates, Tenders, and Contract Changes

GENERAL COMMENTS

Very few projects are constructed without disruptions and delays, but estimators tend to favor the bright side: these problems may not arise on the project under consideration. Even when experience has proven that disruptions and delays do occur, there have often been extenuating circumstances to explain them, and the estimator hopes that lightning will not strike twice in the same place — unfortunately, it often does.

The causes for disruptions and delays are numerous, and most of them can be anticipated. The questions are, how likely is it that they will occur, and, if they do occur, how severe will they be? All too often, estimators will be over-optimistic when answering these questions. When disruptions and delays do occur and the estimate has not provided for them, contractors may want to pass the costs on to others. However, passing the costs on seldom works if it can be shown that the disruptions or the delays, or both, should have been anticipated by the estimator, regardless of their likelihood to occur.

I know of a number of cases involving severe weather conditions that were hard to anticipate. I also know that the contractor involved in each case experienced big losses because of the disruptions and delays, but the construction owners were unwilling even to consider a claim for damages, and were sometimes hard-pressed to allow an extension to the completion dates; instead, they required the contractors to accelerate the work. The same thing can happen with other delays that may be out of the contractor's control but should have been anticipated, even if their likelihood to occur was remote. Often, when the contractor does not anticipate disruptions and delays, the owner considers this oversight as one of the contractor's shortcomings.

The practical implication is this: when disruptions and delays occur that were impossible to anticipate and are clearly not the contractor's fault, they are often intermingled with the contractor's shortcomings and are consequently hard to isolate and hard to get the owner to accept. The owner, who is normally well represented at construction sites, usually keeps his own record of all that happens on site and can often make a good case of the contractor's shortcomings. Therefore, the contractor is well advised to keep his shortcomings to a minimum. Furthermore, there is a practical limit to allowing for all possible disruptions and delays, but the contractor should know the arguments that will face him.

JOB FACTORS

Estimates for construction work are usually adjusted for all kinds of factors that may affect the productivity of the installation labor. These factors are generally referred to as job factors. The estimator must consider making labor adjustments for a great variety of such factors, for example:

- remote geographic locations;
- adverse weather conditions;
- complexity of installation;
- poor design drawings or specifications;
- unknown productivity of strange crews;
- unfamiliar construction, conditions, or work;
- building renovations;
- partially or wholly occupied buildings;
- spread-out work areas (e.g., industrial sites);
- anticipated schedule delays (for whatever reason);
- anticipated work disruptions (for whatever reason).

These are but a few of the job factors that the estimator must consider with every estimate. Wrong evaluations for adjustments will invariably cause labor overruns. Even relatively small estimates, for example, extra work required by an owner, should be adjusted for such job factors.

If the contractor's labor has not been properly adjusted for some obviously required job factors, serious problems can develop when a delay claim arises, since the respondent to the claim will usually look for any and all shortcomings by the contractor in order to reduce the claim costs. (The Appendix provides some details of these problems.)

QUALIFYING TENDERS

Some owners take a dim view of tender qualifications, and contractors should beware of qualifications that could get their tenders rejected. However, some qualifications are absolutely necessary, and contractors would be remiss in their duty to the owner to omit them.

If a contractor becomes aware of problems that could cause serious disruptions and delays to the construction schedule, for example, he should make the owner aware of these problems. Some contractors are reluctant to do so, fearing they may lose the contract. In these cases, hard feelings and conflicts may arise during construction, and the reception of delay claims is usually affected as well. Gaining a contract under these circumstances is seldom profitable, and the contractor is well advised to qualify his tender.

INCLUDING A PRELIMINARY SCHEDULE WITH THE TENDER

Normally, a contractor prepares a construction schedule after entering into a contract, but there is much to be said in favor of preparing a schedule during the tender period — at least on a preliminary basis. Both the owner and the contractor, for different reasons, should be interested in the start and completion dates of construction. Milestone dates for major construction phases are also useful.

If a contractor submits a preliminary construction schedule with his tender, there can be no doubt in the future about his intentions and what his tender was based on. Construction schedules that are submitted after contract awards, sometimes not until many weeks after construction has started, frequently leave a doubt in the owner's mind as to what the tender was based on, especially when schedule changes become necessary. It is critical to remove such doubt to avoid a dispute over the initial intent.

Submitting a preliminary construction schedule with the tender is even more important to a subcontractor than to a general contractor, since a subcontractor is interested in various start and completion dates during the course of construction. However, it is useful for a general contractor to have each subcontractor's input before submitting his own schedule, especially if subcontractors supply materials for critical tasks.

Sometimes a simple schedule that shows a few milestone dates, all critical tasks, and the amount of float time for noncritical tasks is better than one with a complex network of tasks, dates, and dependencies that takes many weeks or months to prepare and an enormous effort to update regularly to be useful. Such a schedule can be hand-drawn, prepared with a computer program such as Microsoft's *Project*, or created in a word-processing program (see Figure 1-1).

Thinking in terms of a construction schedule can also change the way an estimator prepares his take-offs and divides his labor estimate. This division can be useful if manpower loading is to be included with the schedule and as a tool for the project manager in the assessments of the completion status of the project and possible productivity losses. Quite often the value of preparing a simple schedule during the tender period far outweighs the small cost involved to do so.

ASSURING CONTRACT CHANGES INCLUDE IMPACT COSTS

When preparing quotes for contract changes, contractors should consider all possible impact costs and make owners aware of them. These impact costs are often overlooked until later, usually towards the end of construction, when the contractor realizes that their cumulative effect is more than he can absorb. At this time, the owner may not be receptive to a claim for these impacts, and understandably so, because mitigation is no longer possible. A subcontractor should also be aware of other subcontractors' changes, because these changes, too, can have disruptive and delaying effects on his unfinished work and should, therefore, be priced and quoted, if only for their impacts.

TYPES OF IMPACT COST

The impacts caused by change orders to the contracted work are mostly of a disruptive nature and mostly to the work in progress, including previously contracted change orders. In this regard, construction is not too different from assembly lines. If you were to request changes to your car while it is being assembled, the cost of the car would be prohibitive.

Requests for clarification (RFCs) can also disrupt the work while the contractor awaits an answer, although RFCs may *not* end up in contract changes that were priced by the contractor. If this is the case, the contractor should promptly notify the owner's representative and request a contract change, both for schedule delays and impact costs involved. The problem is that, although these disruptions may be insignificant individually, collectively they can cause enormous damage.

Some changes to the contract may require materials with fairly lengthy deliveries that may delay the completion of the entire project; since another trade may cause such a delay, each subcontractor should be afforded the opportunity to evaluate the impact on his work. Other changes to the contract could push summer work into winter, with the effects of adverse weather conditions as the impact. This impact can occur not only to the immediate work affected by the change but also to other parts of the contract. Even worse, the immediate work affected by the change may not be subject to weather conditions but may delay other parts of the work that are thereby pushed into winter.

It is also sometimes necessary to accelerate the work of the project to catch up with the delays caused by change orders, especially if the contract disallows schedule extensions. Accelerations are another type of impact that affect the contractor's costs. Shift work, overtime, increased manpower and supervision, more construction equipment, and additional time to train new workers may be required.

The most deceiving part of change orders is their appearance to be profitable: all direct costs are covered, and overhead and profit margins seem satisfactory. Viewed in isolation, change orders may appear to be acceptable. The contractor seldom looks at the impact they have on other work in progress, including previously issued change orders.

Time Periods

| Tasks | 1 2 3 4 5 6 7 8 9 10 11 12 |

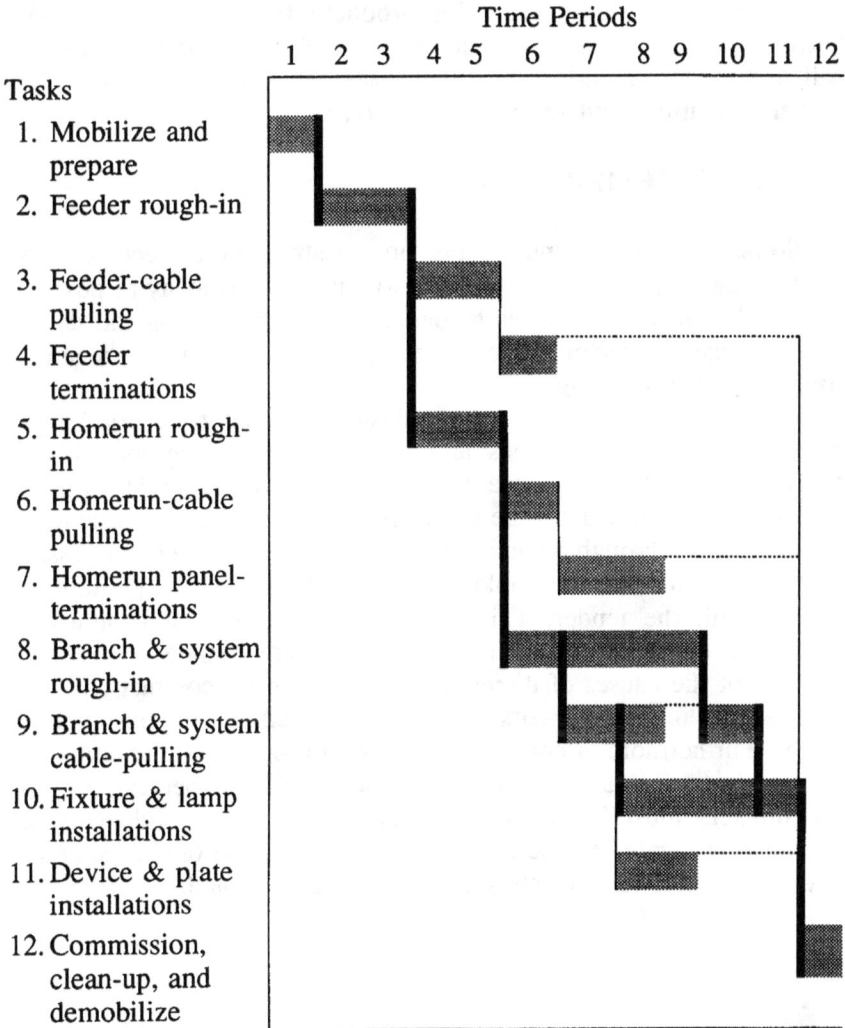

Tasks

1. Mobilize and prepare
2. Feeder rough-in
3. Feeder-cable pulling
4. Feeder terminations
5. Homerun rough-in
6. Homerun-cable pulling
7. Homerun panel-terminations
8. Branch & system rough-in
9. Branch & system cable-pulling
10. Fixture & lamp installations
11. Device & plate installations
12. Commission, clean-up, and demobilize

Typical Construction Schedule for Electrical Work

Figure 1-1

Notes:

The shaded areas represent the required time for each task. Vertical lines indicate task dependencies on previous task(s). Dotted lines represent float, also known as slack, because the tasks connected with them can be "floated", or performed, along these lines without affecting the critical path. Heavy lines plus shaded areas, or partial shaded areas, without float represent the critical path, i.e., tasks along this path are critical because they cannot be delayed without delaying the completion date of the project. Tasks 1, 2, 5, 8, 10, and 12 are critical. Task 8 and part of task 9, and task 10 and part of task 9, form parallel critical paths, i.e., these critical tasks are performed concurrently.

Impact costs involve mostly lost productivity, additional overheads, interest charges, and return on investment, which are treated in more detail in Chapters 2 and 3, but these costs can also include labor and material escalations and extended guarantees.

CHAPTER 1 REVIEW

Some disruptions and delays on construction projects can be expected, and it is up to the contractor to anticipate as many as possible and to make allowances in his tender. Potential disruptions and delays ignored by the contractor at the tendering stage could later be regarded as one of his shortcomings.

A labor estimate is usually adjusted for abnormal conditions, including expected disruptions and delays, by applying job factors. However, if a contractor is aware of circumstances that might seriously affect the construction schedule of the project, it may be advisable to let the owner know through a tender qualification, even at the risk of losing the contract. Another advisable practice is to include a preliminary schedule with the tender. This schedule establishes the contractor's intentions at the outset, which helps resolve future disputes.

One of the causes of disruptions and delays are contract changes, which usually impact all work in progress, including previous contract changes. Furthermore, a contractor's work in progress is also impacted by some contract changes by other contractors. The expected costs for these impacts must be included in quotes for contract changes (or quotation qualifications to supply them later). The value of change orders that exclude costs for impacts should never be overrated.

2

Monitoring Construction Progress, Measuring Productivity Losses, and Recording Construction Constraints

GENERAL COMMENTS

To make a convincing case to be compensated for work delays, disruptions, and slowdowns, a contractor must at least be able to produce a meaningful construction schedule. A construction schedule will, first of all, make the contractor's initial plans known to all concerned. This schedule is usually called "As Planned". Then, as construction proceeds, the schedule is normally updated periodically to show the progress of the work and to make necessary adjustments. When construction is completed, the schedule is usually called "As Built".

Although the schedule can also show average manpower loading for various activities, it is important to realize that it is not designed to measure productivity. A schedule is time related. It shows when activities are planned to start and end, or have started and ended, how long these activities were planned to take, or have taken, and if they were planned to be disrupted, or have, in fact, been disrupted. Even if the average crew size for each activity as planned and as built is shown in the schedule, it is normally not a good indicator of productivity.

Productivity measurements are usually compared with budget expenditures, because the budget assumes 100% productivity; therefore, if an activity costs more to finish than the budget allows, the cost overrun is generally expressed as a percentage of lost productivity, but if the activity takes longer than planned *without* a budget overrun, no reference is made to productivity loss.

Productivity losses are usually accompanied by a loss of time; that is why some people try to establish them through as-built construction schedules. However, quantifications for productivity losses should be established independently of the construction schedule. Methods for measuring productivity losses usually involve some form of planned-to-actual manpower-loading comparison. Some degree of inaccuracy is involved with most methods, but the inaccuracy is minimal in a method referred to as the task-per-area assessment (described in a separate section of this chapter). Regardless of which productivity-assessment method is used, it should complement, rather than substitute for, the construction schedule; but if time allows maintenance of only one of them, it is advisable to choose the task-per-area assessment, because it provides project managers and owners with more timely and, likely, more accurate information, although in less detailed graphic format.

Time Periods

1 → 2 → 3 → 4 → 5 → 6 → 7 → 8 → 9 → 10 → 11 → 12 →

Task #1A (8)

Float

Task #1B (6)

Float

Task #1 (20) | Task #2 (30) | Task #3 (40) Task #4 (34) Task #5 (40)

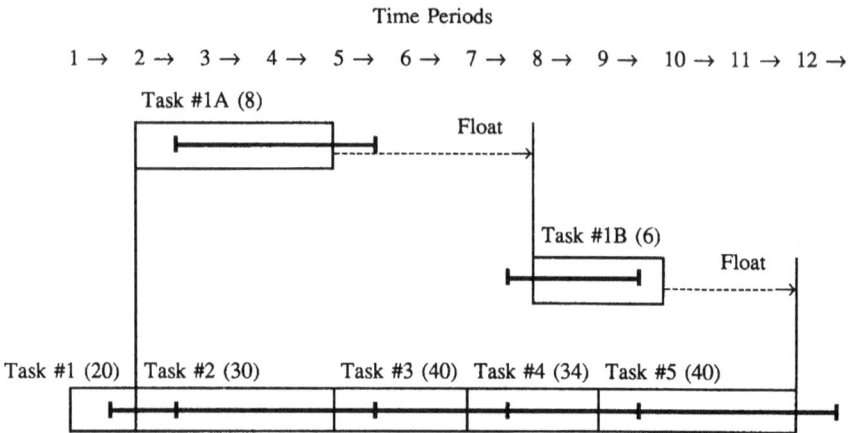

Figure 2-1 Planned and executed times for construction tasks (with an initial delay of a critical task)

THE CONSTRUCTION SCHEDULE

A construction schedule is designed to show the activities necessary for the construction of a project against the time required to construct and complete the project. It usually includes preliminary tasks that are necessary to start construction.

Figure 2-1 shows a simple schedule that suffices for illustration purposes. A full-blown construction schedule is usually much more complex, but is made up of the elements shown here. The time periods shown could be hours, days, weeks, months, or even years. Typically, the periods used are either weeks or months. The rectangles represent the various activities. A vertical line shows the dependency on a previous activity. For example, Task #1A and Task #2 depend on Task #1 to be completed, and Task #1B depends on Task #1A to be completed, and so forth. The dotted lines show float time, which allows noncritical activities to be moved to another time frame. For example, Task #1A is planned to be performed during time periods two, three, and four, but could be performed during time periods five, six, and seven, or even later if Task #1B is also moved. Critical activities have no float time; that is why they are critical. When a critical activity is moved, as shown for the delay of Task #1 in Figure 2-1, all critical activities following it must be moved, and noncritical activities depending on them, such as Task #1A, must be started later. Furthermore, noncritical activities depending on critical activities lose some of their float time.

Time Periods

1 → 2 → 3 → 4 → 5 → 6 → 7 → 8 → 9 → 10 → 11 → 12 →

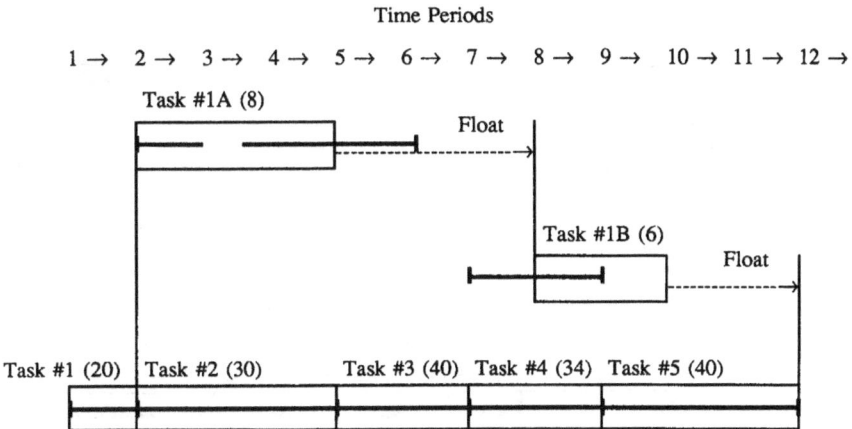

Figure 2-2 Planned and executed times for construction tasks (with a disruption of a noncritical task)

The numbers in parentheses following the task numbers represent the planned *average* workforce required to perform the activity in the time allowed. The actual average workforce required to perform the activity can also be shown below the rectangle, as in Figure 2-5, but in Figures 2-1 to 2-4 inclusive the actual average workforce is the same as the planned one. It is important to realize that this number is the *average*. For example, if the time periods shown in Figure 2-1 represent weeks, then Task #1A would require 24 worker-weeks to complete, and if three workers missed one week each, the Task would require nine workers to maintain the average of eight workers.

Tasks #1 to #5 inclusive are shown in a row for simplified illustration. Normally, these tasks would be vertically staggered to allow for a more detailed description in a left column (see Figure 1-1). The bold lines shown along the centers of the rectangles represent the *actual* (as-built) times when the activities were performed.

A construction schedule is a good means to show time losses, but time losses do not necessarily mean productivity losses and vice versa. For example, Figure 2-1 shows a time loss at the start of the project. It is unlikely that this time loss caused a productivity loss, because the workforce had not yet started. However, a critical activity was involved, and the project will be delayed unless some activities can be accelerated. On the other hand, Figure 2-2 shows a time loss for Task #1A, and this will most likely result in a productivity loss, because the activity was disrupted; but, since the activity was noncritical, float time was used up, and the completion date for the project was not affected.

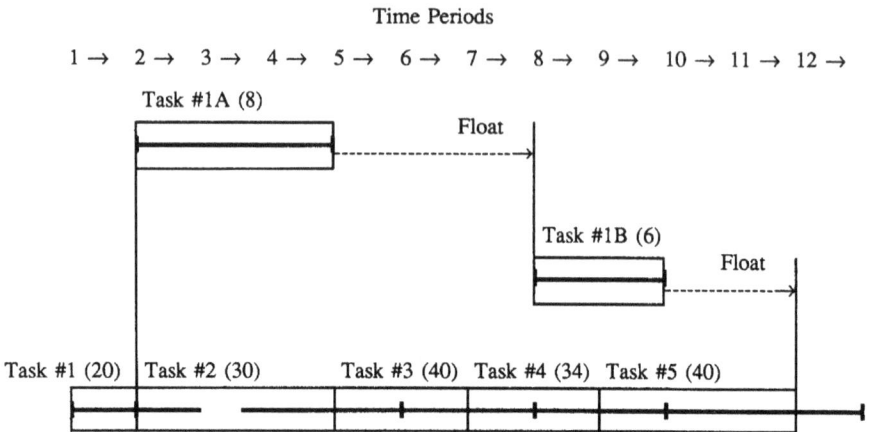

Figure 2-3 Planned and executed times for construction tasks (with a disruption of a critical task)

Figure 2-3 shows a situation, for Task #2, when both time and productivity losses occurred. Since the activity is critical, the time loss caused a project-completion extension. As can be seen, the project-completion extension is longer than the actual time loss: the disruption of Task #2, and its resulting productivity loss, required more than three time periods for the task to be completed.

Figures 2-1 and 2-2 show a loss of float time due to delays of tasks. Sometimes, arguments develop over "who owns float time?" because the owner assumes that, when critical activities are delayed, the contractor is obliged to shift workforces to noncritical activities without a claim for production losses. This is a reasonable assumption, but it does not always work. The contractor may have planned to use noncritical activities to smooth manpower loading (which is possible for workers of the same trade). For example, when Task #3 is completed and Task #4 can only absorb 34 of the 40 workers transferred, 6 workers can be assigned to Task #1B, and when Tasks #4 and #1B are completed, all 40 workers are assigned to Task #5.

On the other hand, if Task #3 is slowed down for lack of information from the owner and six workers have to be temporarily laid off to mitigate the effect, they could be assigned to complete Task #1B. In this case, the company can still claim for a loss, because more workers will have to be hired later to complete Task #5, with a resulting productivity loss due to the required learning curve. Unless the contract provides otherwise, it makes sense that a contractor owns float time if he can show he needs it for smoothing his manpower loading.

Time Periods

1 → 2 → 3 → 4 → 5 → 6 → 7 → 8 → 9 → 10 → 11 → 12 →

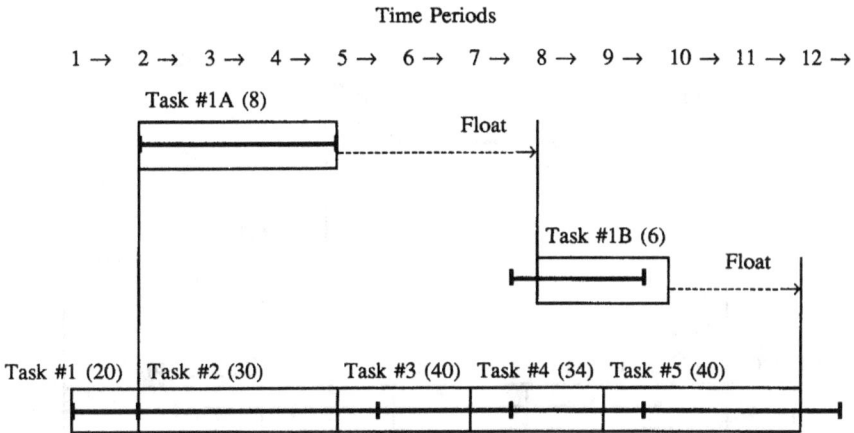

Figure 2-4 Planned and executed times for construction tasks (with a slowdown of a critical task)

A more deluding loss is shown in Figure 2-4. Here, Task #2 was slowed down, but because the crew appeared to be busy, this slowdown was not detected until near the planned completion of the task. As can be seen, the task was completed late and, because the task was critical, it also caused a late completion of the project.

There are numerous reasons for slowdowns: a) late or wrong deliveries of materials; b) inadequate tooling; c) inexperienced or inattentive supervision; d) tardy information supply or approvals; e) changes to the work; f) poor worker morale or attitudes; g) mismanaged materials handling and distribution; h) adverse weather conditions; and so on. Slowdowns always cause delays and production losses. The latter are more severe if the slowdown was deluding, that is, if it was not detected in time by the contractor's supervisory staff to mitigate the situation, which can easily happen when crews appear to be busy; crews try to keep themselves busy in the most adverse circumstances, with expectations, or promises, that improvements are imminent — thus, much unnecessary work is often done to fill the voids left by slowdowns.

With a workforce loaded as planned, even schedule updates usually do not detect slowdowns until it is too late to mitigate them. Schedule updates often occur many days or weeks after the work is done, and are seldom of help while the work occurs. That is why construction schedules are not the best measuring systems for such losses, and why it is so important to look at various construction-progress measuring systems to decide which one can do the best job in the circumstances.

Time Periods

1 → 2 → 3 → 4 → 5 → 6 → 7 → 8 → 9 → 10 → 11 → 12 →

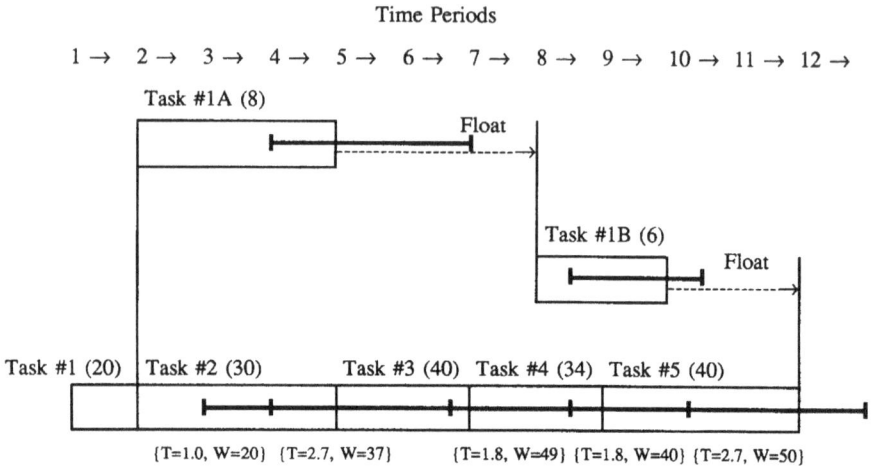

Figure 2-5 Planned and executed times for construction tasks (with an initial delay and later acceleration of critical tasks)

When a schedule is severely delayed, a contractor is normally asked to accelerate the work, as shown in Figure 2-5. The usual way to accomplish acceleration is by increasing the manpower. The numbers in the braces below the bold as-built lines of Tasks #1 to #5 give us the time periods (T) and the average number of workers (W) that were used to perform each task. Since Tasks #1A and #1B have float time, there was no need to accelerate them; however, the contractor was able to accelerate all other tasks except #1.

Work accelerations bring about their own productivity losses, as is evident here, but the whole idea is to attain an earlier project completion date, which is more valuable to the owner than a small loss of productivity. For the critical Tasks #1 to #5, the as-planned worker-time-periods came to 378 (1 x 20 + 3 x 30 + 2 x 40 + 2 x 34 + 3 x 40) and the actual worker-time-periods came to 415.1 (1 x 20 + 2.7 x 37 + 1.8 x 49 + 1.8 x 40 + 2.7 x 50) for a 9% ((415.1 - 378) / 415.1) productivity loss. The project had originally been scheduled to be completed in eleven time periods, and the acceleration was able to reduce the schedule to ten time periods. However, since the project had lost two time periods at the outset, the completion was still delayed by one time period. The value of the gain of one time period (let's say four weeks) should be compared with the cost of the 9% productivity loss. This case may vary with each project; it may also be challenged by the owner, who may doubt that the entire productivity loss was due to acceleration. Besides, if the contractor was responsible for the initial delay, the owner may demand the acceleration without caring about any productivity loss.

Productivity losses are highly subjective, as we shall see in the next section, and the establishment of one or more methods for measuring productivity, regardless of whether or not delay claims are expected, is desirable.

METHODS OF MEASURING PRODUCTIVITY

Delay claims usually consist of productivity losses and various additional overheads. Establishing overheads can be fairly complicated (see Chapter 3) but this is simple compared with establishing productivity losses. Productivity losses are harder to establish because they occur even during so-called normal periods and usually overlap with productivity losses caused by abnormal disruptions and delays. Tying productivity losses to specific disruptions and delays can often prove impossible; nevertheless, some methods are successful at providing fairly close assessments and deserve our attention.

There are six main methods of measuring productivity and productivity losses:

1. Manpower-loading Comparison;
2. Differential Measurement, also known as the "Measured Mile";
3. Billing-to-Cost Comparison of Various Billing Periods;
4. Snapshot Technique;
5. Task-per-Area Assessment Comparison to Budget Expenditure;
6. Weekly Constraints Assessment Report.

Each of these methods offers advantages and disadvantages, as we shall see in the following subsections.

Manpower-loading Comparison: This method compares the average as-planned with the average as-built durations and manpower loadings for each task. For example, Task #2 in Figure 2-4 was planned to be accomplished by an average of 30 workers over 3 time periods; it actually took an average of 30 workers 3.6 time periods to complete this task. The productivity loss of 30 workers for 0.6 time periods obviously has some cause that must be carefully assessed. However, this assessment may be difficult because, over this length of time, several loss causes may have occurred, and there may have been a substantial time lag before the schedule was updated and the overall loss was discovered. Furthermore, the cause(s) may have been deluding, that is, hardly noticed by anyone during this period. The method is more accurate if a definite delay or disruption cause is present, as for Task #1 in Figure 2-1 or for Task #1A in Figure 2-2. Also, when accelerations are planned and carried out, this is probably the best method to make a good case for productivity losses (see Figure 2-5). However, this method is still subjective, since other productivity losses that had absolutely nothing to do with the acceleration could have occurred at the same time.

Differential Measurement or "Measured Mile": This method compares an abnormal (impacted) period with a previously established normal (unimpacted) period. The so-called normal period would include any and all estimating and/or operational mistakes made by the contractor, and the abnormal period would include the identifiable claim cause(s). For example, in Figure 2-2 the Task #1A time period before the disruption would be considered normal and the time period after the disruption, abnormal. Thus, the duration and manpower required per percent of work done in the normal period compared with the abnormal period establishes the differential, or measured mile, for the lost productivity. There are, however, several objections to this method.

For one thing, the so-called normal time period for Task #1A in Figure 2-2 is very much affected by the learning curve (described in more detail in the next section). For another, consider Task #2 in Figure 2-4, or Tasks #2 to #5 in Figure 2-5: What is the normal period? It cannot be Task #1 because Task #1 consists of completely different work; for a comparison to be valid, the work must be identical, and the working conditions, work forces, and so on, must closely match. The problem with this method is that construction usually consists of a multiplicity of different tasks, carried out under different circumstances and with different crews, which make comparisons almost meaningless.

Billing-to-Cost Comparison of Various Billing Periods: This method is similar to the measured mile but uses billings to establish the percentage of project completion during the normal period of construction. It also assumes that costs of construction are in line with billings during the normal period; therefore, any labor and overhead costs that exceed the billings during the impacted period of construction should be claimable. In addition to the objections already mentioned under Differential Measurement, there are a few more to be considered for this method, such as stockpiled materials and unearned revenue (overbillings).

I once had an opportunity to observe a claim preparation for a project that consisted of installing pipe[*] for three-quarters of the construction period and fixtures for the last quarter. The second and third quarters were severely impacted by owner-caused disruptions and delays. However, at the end of construction the contractor could not produce any evidence of progressive productivity measurements, and the hired claims consultant decided to use the contractor's billing record to prove the

---

[*] This example could apply equally to mechanical and electrical work.

effects of the impacted construction period. He used the first-quarter billings for the percentage of work done during the normal period of construction and compared this percentage with that of the next two quarter billings, being the impacted period of construction. However, this produced an overrun percentage that was unreasonably high.

As I had been asked (by the contractor) to observe and point out flaws, I pointed out that the first-quarter billings were not only overbilled but also included a substantial stockpile of materials that had not yet been installed. The claims consultant readily adjusted his calculations to discount these factors, but the percentage productivity loss was still too high, in my opinion. I interviewed some of the workers who were involved in the first-quarter of construction to find out what was actually installed during this period and was told that mostly the larger-sized feeder lines had been installed. These feeder lines consisted of long, unobstructed runs, compared with the much shorter and obstructed runs that were installed during the second and third quarters of construction. Also, the material-to-labor-cost ratio was much higher during the first quarter, due to the larger pipe sizes.

I pointed all this out to the contractor and his claims consultant, but they refused to make any further adjustments, since the second calculation neatly fitted the cost overrun that the contractor wanted to recover. However, the owner also detected the flaw and accused the contractor of including his own shortcomings in the claim. In the end, the contractor had to settle for far less than he would have been entitled to with a properly adjusted claim. With the numerous flaws inherent in this method, I would be reluctant to recommend it, unless no other method were practicable.

Snapshot Technique: This method takes stock (a snapshot) at various points in time to determine the amount of time still required to complete the project. For the first snapshot, if the time required to complete the project exceeds the planned completion date, the work done since the planned start date of the project has been delayed and should be analyzed for delay cause(s). For subsequent snapshots, if the time required to complete the project exceeds the completion date established at the previous snapshot, there is a further delay that must be examined for delay cause(s); if the time required to complete the project has advanced, the contractor was able to accelerate the work — perhaps at some loss of productivity that may be claimable.

The best time to take a snapshot is at the completion of a task, because at other times it is almost impossible to establish the time still required to complete the project. For example, if a snapshot was taken during the performance period of Task #2 in Figure 2-4, it would be extremely hard, if not impossible, to establish the time still required to complete the project, but if the snapshot is taken at the completion of the task, it would be fairly easy to establish. Besides, because of the productivity-gain effect of the learning curve, the production average is more meaningful at completion than at any time during the performance of the task.

Obviously, this method works only for critical tasks. For example, a snapshot taken at the completion of Task #1A in Figure 2-2, which is noncritical, would only establish the delay of Task #1A but not any possible delay to the completion of the project. This is one of the flaws of this method. Another flaw is this: since it is best to take a snapshot at the completion of a task, it may become extremely hard to establish the cause(s) of delay incurred during a lengthy task construction period. Furthermore, it may also be too late to mitigate the damages or to give the required notice(s) under the contract to the responsible party.

Task-per-Area Assessment Comparison to Budget Expenditure: This method, too, takes snapshots, but of progressive project-completion percentages on a more regular basis, usually weekly to coincide with the payroll. Regardless of whether or not claims are involved, this is a valuable method for the contractor to determine fairly accurately the progress of project completion. After all, if his labor estimate was wrong, or if his workers are under-performing, or if his material suppliers are letting him down, he must put himself in a position to take prompt remedial actions, and this method is the best I know of to do that. Every contractor should use it on every project.

This method is also used by some owners who wish to monitor the status of completion of a project. In fact, I have met owners who had better knowledge about the project's status than did the contractors whom they employed. Furthermore, some contracts, for example unit-price contracts, use this method to monitor, and pay for, installed work-units.

There are a number of advantages to this method:

1. Since the completion percentage is based on the percentage of the work that must be installed in each area rather than the percentage of labor that was estimated for each area, the influence of a wrong labor estimate is automatically eliminated; for example, if fifty fixtures must be installed in an area and twenty have been installed, the task is 40% completed;
2. More frequent, smaller assessments yield more accurate results;
3. Through frequent assessments, delays are detected more quickly;
4. It makes possible timely notices to responsible parties; and
5. It facilitates prompt remedial actions.

The productivity level is determined by comparing the project completion percentages with the labor budget expenditure percentages. This method is explained in more detail later in this chapter.

Weekly Constraints Assessment Report: This report is prepared as part of the Weekly Field Progress Report, which is explained later in this chapter. This report is extremely useful, especially if completed in conjunction with the Task-per-Area Assessment, because it makes timely and accurate assessments of situations that develop every week. When the results of these assessments are compared with the results of the Task-per-Area Assessments, a compelling argument can be made for the cause and effect of each disruption and delay. Each assessment takes into account the number of workers involved and the amount of time lost due to the constraint, together with appropriate explanations of cause and effect as well as reasons for assigning the responsibility for the damages incurred. A copy of this report can be sent to the alleged responsible party, but care should be taken that it assumes neither proper notification under the contract nor the final estimate for the damages incurred. To avoid the appearance of personal bias, field supervisors must be fair and truthful in their assessments.

THE LEARNING CURVE

At commencement or recommencement of a task, there is a period of low productivity, which gradually improves until it reaches an optimum level. This is known as the learning curve. The lower productivity is due to unfamiliarity with the product or the process, and to the required planning, mobilization, and setup time at the outset. The learning curve can be fairly flat, for simple tasks, or fairly steep, for complex ones, but there is always a learning curve. This fact is important to keep in mind when conducting productivity measurements. It is also important to understand the makeup of a learning curve and how it affects the average rate of production.

A learning curve is usually portrayed as a continuously declining graph, eventually levelling out, similar to the one shown in Figure 2-6. However, this graph is, at best, the average of a series of shorter graphs with regular disruptions, similar to those shown in Figure 2-7. Normally, this difference would not matter because the average should suffice for most calculations, but our aim is to analyze claim causes, and disruptions are at the top of the list. It can be unequivocally stated that disruptions, even minor disruptions, cause a loss of productivity, after which the learning curve restarts at a higher point. In the case of Figure 2-7, the illustrated disruptions represent daily (DB) and weekly (WB) breaks at

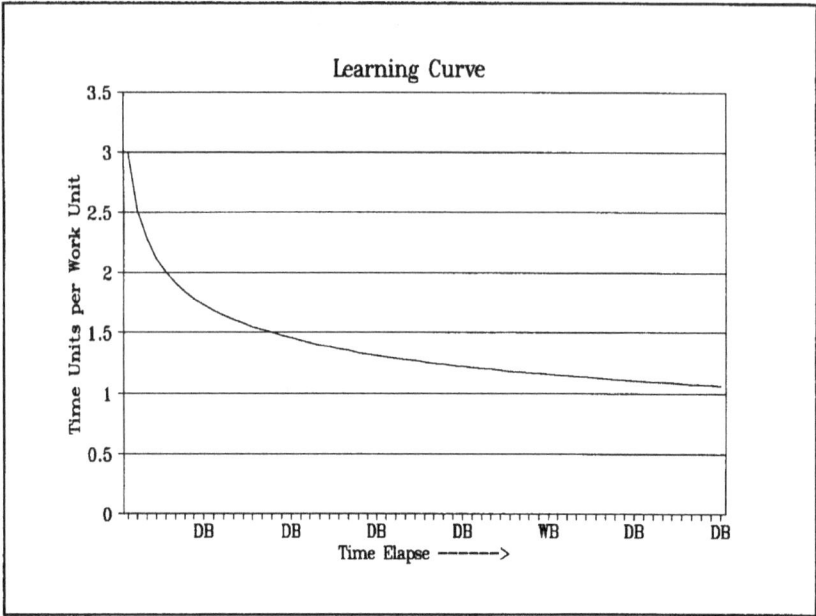

Figure 2-6 Averaged production

the end of shifts. There is usually a slight loss of productivity just before the disruption because of demobilizing procedures. Longer disruptions, such as employee vacations, have a worse effect, and crew turnover has the worst effect, because it tends to restart the learning curve. That is why contractors are often reluctant to lay off workers during disruptions.

When the learning curve is replotted to represent achieved productivity, it looks a little like a sawtooth: it usually starts substantially below the 100% productivity level and ends up slightly above it; thus, with each succeeding task, several sawtooth curves will occur (see Figure 2-8, graph bp). The average of this graph, however, is equal to 100% productivity *if* the labor estimate is correct and *if* nothing occurred to change it. Furthermore, when several tasks on one project are being performed simultaneously, the sawtooth effect will all but disappear, and the overall productivity graph may end up fairly level.

The sawtooth effect of productivity influences all measurements, but less so in the case of Task-per-Area Assessments, because this method breaks the project down to a larger number of areas in which the various tasks are being performed, and simultaneous learning curves occur as a matter of course. Therefore, and for ease of illustration, only the straight-line averages for productivity are shown in the graphs of the next section on establishing Task-per-Area Assessments.

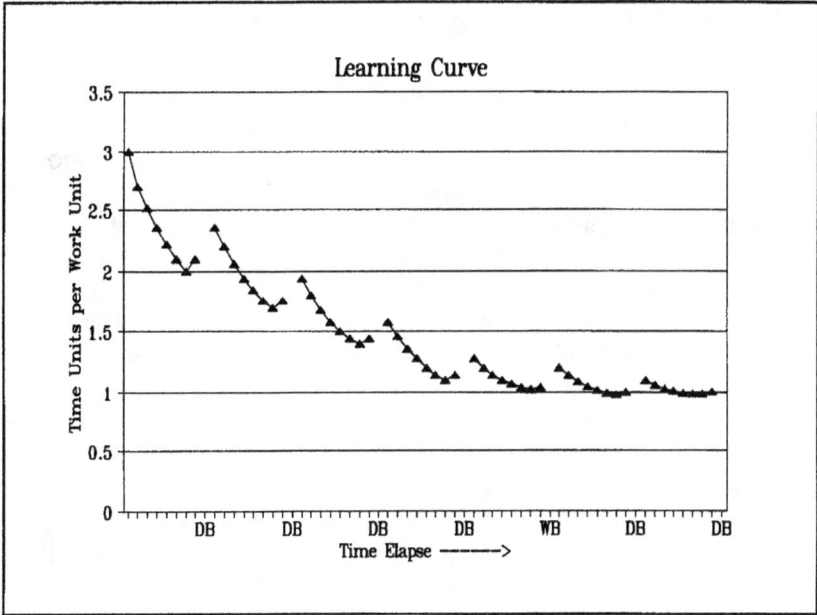

Figure 2-7 Close to actual production

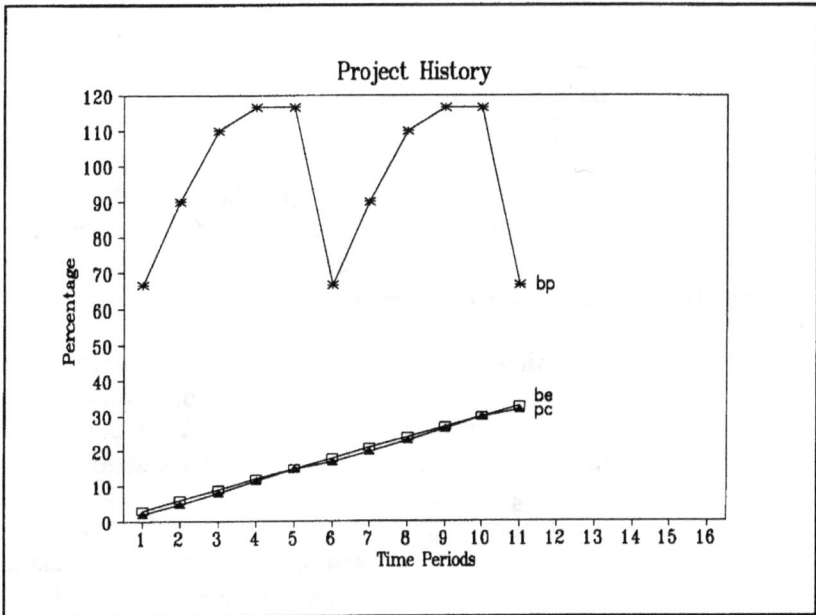

Figure 2-8 Labor-cost history (with the learning-curve effect)

ESTABLISHING THE TASK-PER-AREA ASSESSMENT

As we have seen in an earlier section, construction schedules are useful for graphic illustrations of timeliness and interdependencies of the various tasks, but they do not provide reliable identification and measurement of specific productivity losses. A good example of this unreliability was shown in Figure 2-4, where the actual manpower loading matched the planned manpower loading for Task #2, but the task took longer to complete. A productivity loss could have occurred at any time or at all times during the performance of this task. With schedule updates lagging behind construction, the task could be near completion, or could even be completed, before project management realizes that something went wrong during its performance. By then, it may even be too late to know exactly what went wrong — never mind taking any corrective actions or giving timely notice to the guilty party. It is important, therefore, to use a measuring system that detects productivity losses promptly, preferably weekly or coinciding with the payroll.

A measuring system for progressive project completion must be simple to set up, simple to update, independent of the construction estimate, and fairly accurate. I have used such a system many times, and it has served me well during my years in construction. The complete system includes an assessment of the completion of each task per area and a comparison of the total installation completion percentage with the percentage of labor-budget expenditure.

To set up the task-per-area assessment, one must first determine the various tasks to be performed and how many areas should be chosen for each task-completion assessment. The number of tasks varies with each trade, and the number of areas depends on the desired detail and accuracy. Fewer areas also mean fewer assessments, but the assessments are more difficult, with a corresponding loss of accuracy. Three tasks and five areas were chosen to illustrate the concepts in this section (see Table 2-1).

The second determination should be the amount of work that has to be done for each task in each area. The construction estimate can be used as a guide to establish this determination, but the final judgment call should come from the field supervisor, because the construction estimate is largely based on averages and may incorporate some mistakes. The amount of work per task per area is then compared with the total amount of work of *all* tasks in *all* areas; thus, each ratio represents the weight of its portion of the total installation, and the weights of all tasks in all areas total 1.0. For example, in Table 2-2, Task #1 in Area #1 has a weight of 5/100; that is, it represents 5% of the total installation. The denominator of 100 was chosen for ease of illustration; in practice, this

Work Areas	Tasks			
	1	2	3	All
1				
2				
3				
4				
5				
All				

Table 2-1   Task-per-area completion (blank form)

Work Areas	Tasks			
	1	2	3	All
1	5/100	3/100	6/100	14/100
2	6/100	4/100	7/100	17/100
3	7/100	5/100	8/100	20/100
4	8/100	6/100	9/100	23/100
5	9/100	7/100	10/100	26/100
All	35/100	25/100	40/100	100/100

Table 2-2   Task-per-area completion (time used)

denominator could be any number related to the total work, either expressed in dollar cost or in worker hours; if dollar cost is used, be careful to allow for inflationary increases that could increase the worker-hour costs in later construction periods. The weight for each task in each area becomes part of a hidden formula in each cell of an electronic spreadsheet used later to determine the total installation-completion percentage.

The third determination comes from the field supervisor, and is established at regular intervals, preferably weekly. The field supervisor determines the percentage of completion of each task in each area and fills in his report to project management accordingly (see Table 2-3). To accomplish this, the field supervisor counts the number of work units of

Work Areas	Tasks			
	1	2	3	All
1	60%	33%	17%	
2	50%	25%	14%	
3	43%	20%	25%	
4	25%	33%	33%	
5	33%	43%	30%	
All				

Table 2-3   Task-per-area completion (individual percentages)

Work Areas	Tasks			
	1	2	3	All
1	60%	33%	17%	35%
2	50%	25%	14%	30%
3	43%	20%	25%	30%
4	25%	33%	33%	30%
5	33%	43%	30%	35%
All	40%	32%	25%	32%

Table 2-4   Task-per-area completion (all percentages)

each task installed in each area and compares these installed work units to the total number of work units that must be installed in this area. For example, if 80 fixtures must be installed in an area and 48 have been installed, the task is 60% complete in that area; furthermore, if this is Task #1 in Area #1, which had a total installation weight of 5% (see Table 2-2), 60% completion of this task in this area represents 3% of the total installation (0.60 x 0.05 x 100).

The next determination is the calculation of the total installation-completion percentage, which is done in the project management office with the help of an electronic spreadsheet. The field supervisor's task-per-area completion percentages (from Table 2-3) are entered in the electronic spreadsheet, and the formulas in this spreadsheet generate the completion percentages for *all* tasks and *all* areas (see Table 2-4). Thus, Table 2-4 shows us that Task #1 in all areas is 40% complete, and Work Area #1 including all tasks is 35% complete. Similarly, all tasks in all areas (see bottom right cell) are 32% complete. This total installation-completion percentage is then compared with the percentage of expended labor budget to determine if the work period incurred a productivity loss or gain.

It is important to use a consistent assessment method and avoid weekly mind changes when assessing the completion of each task; that is, if installed work units are compared with total work units one week, do not start guessing at the completion percentages in the next week. Inconsistencies from one week to the next will show up very quickly in a productivity graph, which normally shows only external influences.

To make the installation assessments more accurate, you would be wise to separate tasks such as corrective work, incidental work (such as testing and commissioning), and field supervision (if included with the labor budget) from the main installation tasks and create individual tasks for them; thus, the installation, per se, is 100% complete when all installation units are installed.

One last point is that contract changes may affect the weights of particular tasks in particular areas. If this is the case, one or more numerators and all denominators should be changed, keeping in mind that all weights must total 1.0.

TRACKING JOB PERFORMANCE AND PRODUCTIVITY LOSS

Ideally, the total project-completion percentage (pc) equals the budget-expenditure percentage (be), as shown for the first ten time periods in Figure 2-9, which results in 100% productivity (bp). If the project-completion percentage drops even slightly below the expended-budget percentage, there is a drastic drop in productivity (see time period 11 in Figure 2-9). There may have already been indications of this productivity loss at the job site, but the tracking results would initiate immediate investigations, remedial actions, and notification to the party who caused the problem, if there is such a party.

Ideally! But budgets are based on estimates, which are based on estimated averages. Project-completion percentages can be below budget-expenditure percentages if the budget is high (obe), or above these percentages if the budget is low (ube), in which case the productivity percentages are at "ubp" or at "obp", respectively (see Figure 2-10). Nevertheless, the productivity percentage stays even, at either the higher or the lower level, unless the weighting for the tasks is wrong or unless some external influence causes a change in productivity, as shown for

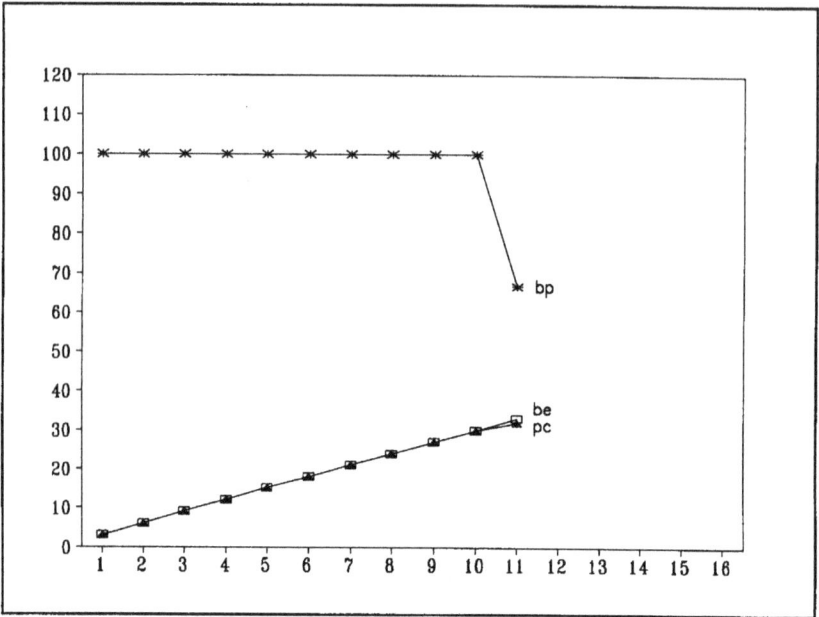

Figure 2-9 Labor-cost history (with a sudden productivity loss)

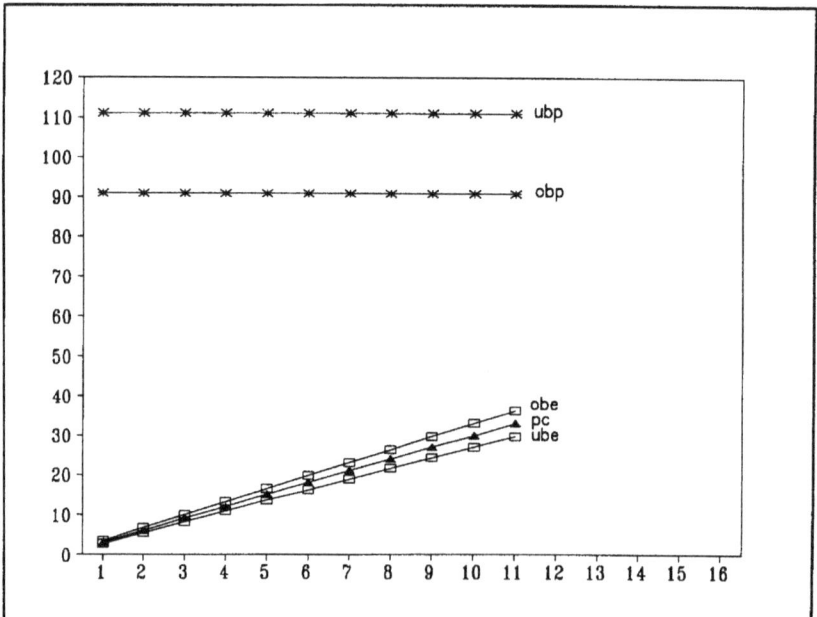

Figure 2-10 Labor-cost history (with over- and underestimated labor costs)

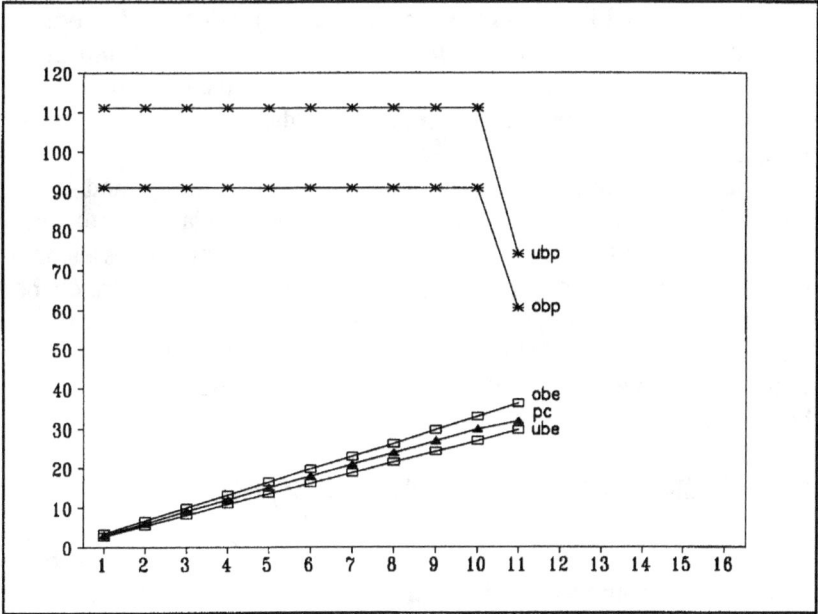

Figure 2-11 Labor-cost history (with over- and underestimated labor costs plus the sudden productivity-loss effect on each)

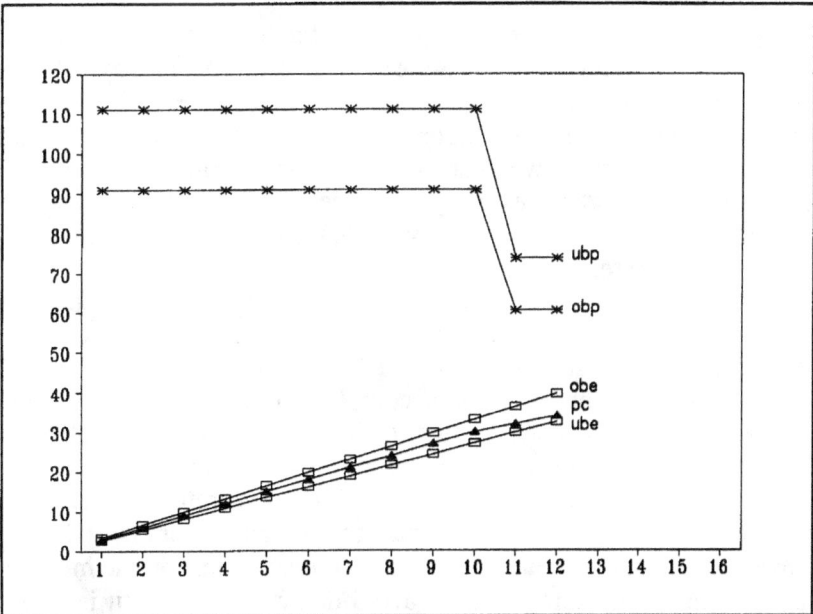

Figure 2-12 Labor-cost history (ditto, but continuing productivity-loss effect)

time period 11 in Figure 2-11. Once the cause for the productivity loss is corrected, the productivity percentage (ubp or obp) will return to its previous level, but if no corrective actions were taken or possible, the productivity percentage will continue at the dropped level, as shown in Figure 2-12.

The big advantage of this productivity measuring and tracking method is that results are made immediately available to project management, and regardless of the causes for productivity losses, that is, whether external or internal, project management is in a much better position to take remedial actions. Also, after analyzing the field supervisor's Weekly Field Progress Report (see following section), project management may wish to give an appropriate notice to the party who is causing the disruption and delay.

THE WEEKLY FIELD PROGRESS REPORT

The field workers and supervisors are, in my opinion, in the best position to judge and assess the status of productivity. Many times I went to jobsites and asked, "How is it going?" and I would get answers like "Very well!" along with a proud description of their accomplishments, or "Not so good!" and I would ask, "What's wrong?" and would get an exact description of the cause and effect of some problem that slowed them down. At such a moment, when everything is fresh in the mind, the field workers and supervisors are probably the best source to assess their productivity loss and to pin it to the right cause. This is why a Weekly Field Progress Report is so important. Nevertheless, the Weekly Field Progress Report is the summation of, not a substitution for, the daily diary, which records weather, events, and employed construction equipment and workforces in greater detail. Every field supervisor, regardless of the size of the project, should be required to keep a consistent, accurate, and updated daily diary.

The very simplicity of the Weekly Field Progress Report speaks for itself (see Figure 2-13), but its enormous advantages go far beyond that: it also provides an effective means of "management by objectives", for example, "here is what we have accomplished in the past week, here is what held us up, and here is what we are planning to accomplish next week". The Constraints section is the most important one in this report because it provides the early warning signal that remedial actions are necessary for damage control. The parties responsible for disruptions and delays must be notified, and all concerned must partake in the mitigation process. This is the surest way to avoid delay claims, and it is also the best way to establish causes and effects of disruptions and delays, which is so important if delay claims become unavoidable.

XYZ CONSTRUCTORS
Weekly Field Progress Report #_____

Project:_____   **Date:**_____

Work Performed During Past Week:

Average # of Workers Employed:_____   Workers Required per Schedule:_____

Constraints:

Time Lost (Manhours):_____   Job Schedule: A) On Time:_____

                                   B) Late By (Calendar Days):_____

Estimated Cost: $_____   C) Ahead By (Calendar Days):_____

Work Planned For Next Week:

Required Workers:

Required Materials:

Required Equipment:

Supervisor's Signature:_____   **Copy To:**_____

Figure 2-13 Weekly field progress report

It is my strongest recommendation to every contractor to establish the Weekly Field Progress Report on all projects from the outset and instruct the supervisory staff on the proper completion of it. The advantages of this report will soon become apparent, especially when combined with the Task-per-Area Assessment. Everybody's efforts to get the project finished on time and within budget is much more focused. The Weekly Field Progress Report can also become evidence if any delay claim is pursued to trial or arbitration; it is the memory of the maker recorded contemporaneously with the events.

Furthermore, notwithstanding the provisions of the contract with respect to notifications, which should be followed carefully, the Weekly Field Progress Report is a handy method to inform the general contractor, or the owner, as the case may be, of weekly planning, activities, and constraints.

CHAPTER 2 REVIEW

In this chapter, we covered variously developed construction schedules and how they affect the project's completion date. We also examined six methods of measuring or assessing productivity and their advantages and disadvantages. Then we dealt with learning curves, learning curves within learning curves, and their effect on production. The section "Establishing the Task-per-Area Assessment" described a recommended method for determining progressive percentages of project completion. We then dealt with how these percentages of project completion are compared with the percentages of budget expenditures to determine variations in productivity. Finally, we covered the Weekly Field Progress Report, which is a valuable tool to speed up the information flow, especially with respect to constraints. This report is also useful for prompt assessments of claim causes and effects, and for initiating prompt remedial actions. If delay claims are to be avoided, the Weekly Field Progress Report is probably the best means to accomplish this task.

3

Allocations of Overheads,
Risk Allowances, Interest, and
Return on Investment

THE DIVISION OF OVERHEADS

Overheads are costs to the company just as materials and labor required to erect construction projects are costs. No claim is properly compiled without overhead costs. In construction, there are three types of overhead costs that must be considered:

1. The costs of maintaining a company's home office and marketing its services.
2. The costs of running the projects that a company has obtained through its marketing efforts.
3. The costs of financing a company's projects.

The first category is known by the terms "general overhead", "company-related overhead", or simply "company overhead", which exists without any of the projects and is, therefore, marketing related; in other words, it is the overhead, including the estimating costs, that the company requires to establish its revenue. Nevertheless, it is an overhead that must eventually be paid for by all the projects that the company obtains; therefore, a proper allocation of this overhead must be assigned to each project.

The second category is known as "project-related overhead" or simply "project overhead". It includes all direct job costs other than materials, labor, and subcontracts. These costs include project management and installation supervision, tools and construction equipment, office and storage trailers, and so on; the costs would not exist without the project and are, therefore, much easier to account for. An allocation is usually not required unless the item cost is shared, such as a tool that is bought for all projects.

The third category arises because of the time lags between paying for a project's expenses and collecting on its billings. Billings are submitted at the end of a month for payment thirty days later, and prior to receiving payment, a contractor must finance eight payrolls (for weekly pay) and materials for at least fifteen days, if he pays for them on the fifteenth of the month following their receipt. Then, there is usually a holdback on the billings that is held until forty-five days after project completion. Together, the financing cost can easily exceed 1% of the total project cost, even if payments are made on time.

Company budgets are not only helpful in understanding the various overhead relationships, but can also be a tremendous aid for understanding where the company is headed and what needs to be done to get there. Furthermore, budgets are invaluable when claims arise. Table 3-1 shows a typical budget format.

Company Budget for the Year 2003
(Revenue Expectation = $2.5 Million)

Direct Labor & Material Costs (incl. Subcontracts)	
2 Projects @ $600,000 Total Base Cost	$1,200,000.00
5 Projects @ 90,000 Total Base Cost	450,000.00
7 Projects @ 50,000 Total Base Cost	350,000.00
Total	$2,000,000.00
Company's General Overhead	
Manager (incl. car)	$75,000.00
Estimator	48,000.00
Assistant Estimator	32,000.00
Bookkeeper/Secretary	24,000.00
Rent (incl. utilities)	10,000.00
Telephones & Long Distance	3,000.00
Supplies & Depreciation	8,000.00
Total	$200,000.00
Company's Project Overhead	
Supervisor (incl. car)	$66,400.00
Expediter	45,000.00
Clerk/Typist	23,300.00
Tools, etc.	23,800.00
Site Offices	17,500.00
Storage Facilities	14,000.00
Misc. Deliveries	7,000.00
Payroll Services	3,000.00
Total	$200,000.00
Other Allowances	
Interest on Loans	$25,000.00
Known-risks Allowance	25,000.00
Profit*, including Unknown-risks Allowance	50,000.00
Total	$100,000.00
Budget Revenue	**$2,500,000.00**

Table 3-1

---

* To provide a fair return on the owners' (shareholders') investment of $250,000.

THE DETERMINATION OF PROJECT OVERHEAD

Even though project overhead (POH) is easy to account for during construction, it is only an estimate during the tender period. For the tender, it can be estimated by assessing each item of cost individually or by using average percentages for material-related and labor-related costs, as shown in the graph of Figure 3-1. The percentages[*] are for the various total base costs (TBCs) and are higher for labor and lower for bigger volumes. Total base costs include both direct labor and material costs. Direct material costs include subcontracts. The markups applicable for a total base cost of $500,000, for example, would be 6% on the material base cost (MBC) and 15% on the labor base cost (LBC).

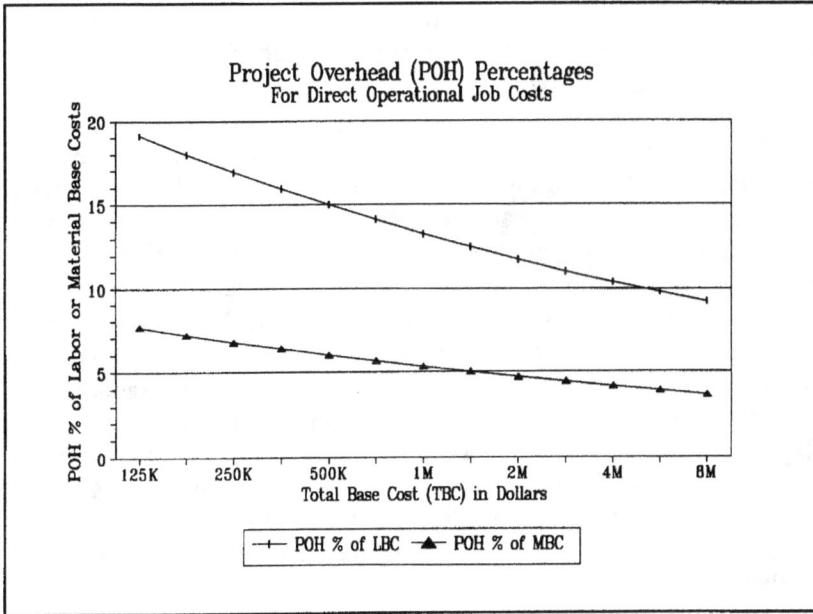

Figure 3-1

---

[*] For those who prefer calculations in lieu of graphs, the following equations can be used (where TBC = total base cost):

$$Material\text{-}related\ Overhead\ \% = \frac{60}{TBC^{(1/5.7)}}$$

$$Labor\text{-}related\ Overhead\ \% = \frac{150}{TBC^{(1/5.7)}}$$

Thus, if the TBC of $500,000 is made up of $400,000 MBC and $100,000 LBC, the estimator would allow 6% of $400,000 plus 15% of $100,000 for his project overhead. Therefore, the total project-related overhead, without adjustments for efficiency or anticipated project delays, would come to $39,000 (0.06 x $400,000 + 0.15 x $100,000).

However, if the estimator prefers to calculate the items for his project overhead separately, the calculations may look as follows:

Supervisor (26 wks @ 12 hrs/wk @ $37.50/hr)	$11,700.00
Expediter (26 wks @ 12 hrs/wk @ $31.50/hr)	9,828.00
Secretary (26 wks @ 12 hrs/wk @ $15.00/hr)	4,680.00
Tools, etc. (est. based on 4% of Labor)	4,000.00
Site office (6 mo @ $630.00/mo)	3,780.00
Site storage (6 mo @ $500.00/mo)	3,000.00
Postage & misc. deliveries (est.)	1,650.00
Payroll service (est. based on 0.5% of Labor)	500.00
Total project overhead	$39,138.00

Remember, neither the average percentage used per Figure 3-1 nor the detailed calculation above is more than an estimate. Both may require adjustments for efficiency and for possible delays. The efficiency may not necessarily involve people efficiency but may be limited to marketing efficiency, if the company does not obtain its budgeted revenue.

Since the personnel for the yearly project overhead must be in place at the beginning of the year to service the projects as they come along, any shortage of budgeted revenue will cause an inefficiency in the budgeted project overhead. A similar situation arises when projects are delayed. For delayed projects, a portion of the project's revenue is either pushed into the next fiscal year or, if the delayed revenue remains in the same fiscal year, it is serviced longer by the project's personnel at the expense of some of the company's undelayed revenue -- an inefficiency in either case.

ALLOCATING GENERAL OVERHEAD

Project overhead can be estimated; general overhead must be allocated. If a company uses a good cost accounting system, the items of project overhead can be accurately established — even for the delay period of a delayed project — but general overhead must still be allocated, not only for the estimate but even for a completed project. This allocation is usually a judgment call by company management. As far as claims are concerned, there is still disagreement over the proper allocation of general overhead, but the US courts and arbitration boards have widely accepted a method (since 1960) known as the Eichleay Formula.

THE EICHLEAY FORMULA

This method uses three steps to arrive at the allocated overhead for a delay period:

Step 1:   A x B / C = D

Step 2:   D / E = F

Step 3:   F x G = H

Where:
   A = The total revenue from the delayed project.

   B = The total company overhead during the delayed project's construction period.

   C = The company's total revenue during the delayed project's construction period.

   D = The company overhead allocable to the delayed project.

   E = The total days of construction of the delayed project.

   F = The daily company overhead allocable to the delayed project.

   G = The number of days the delayed project was delayed.

   H = The amount claimable for unabsorbed company overhead.

The Eichleay Formula tries to accomplish two things: first, it tries to establish, in Step 1, the company overhead allocable to the entire delayed project; it accomplishes this by taking the ratio of the delayed project's revenue to the total company's revenue during the total construction period of the delayed project and applying this ratio to the company's general overhead during the same period.

Second, the Eichleay Formula, in Steps 2 and 3, establishes the daily overhead by dividing the allocable overhead for the total project by the number of days for its construction, and then multiplies this daily overhead by the number of days the project was delayed to arrive at the amount claimable for the unabsorbed overhead.

EXAMPLE OF HOW THE EICHLEAY FORMULA WORKS

The results of the Eichleay Formula can best be demonstrated by using the illustrations below.

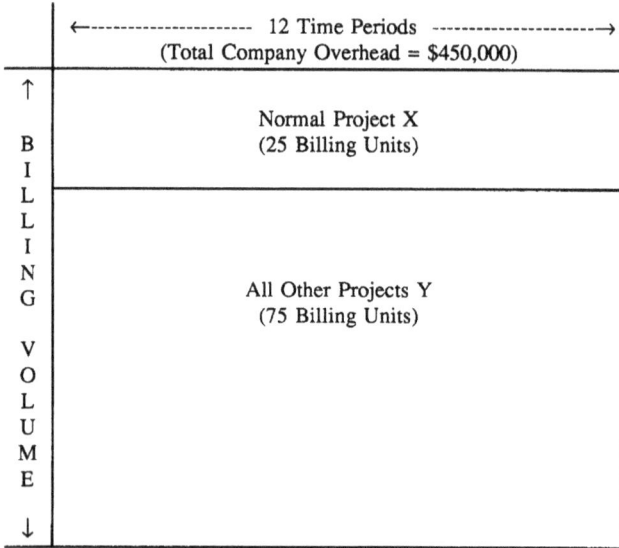

```
        ←------------------- 12 Time Periods -------------------→
                  (Total Company Overhead = $450,000)

  ↑     ┌────────────────────────────────────────────────────┐
        │                                                     │
  B     │                  Normal Project X                   │
  I     │                  (25 Billing Units)                 │
  L     ├────────────────────────────────────────────────────┤
  L     │                                                     │
  I     │                                                     │
  N     │                                                     │
  G     │                All Other Projects Y                 │
        │                 (75 Billing Units)                  │
  V     │                                                     │
  O     │                                                     │
  L     │                                                     │
  U     │                                                     │
  M     │                                                     │
  E     │                                                     │
  ↓     └────────────────────────────────────────────────────┘
```

Figure 3-2   Time periods for company overhead and revenue

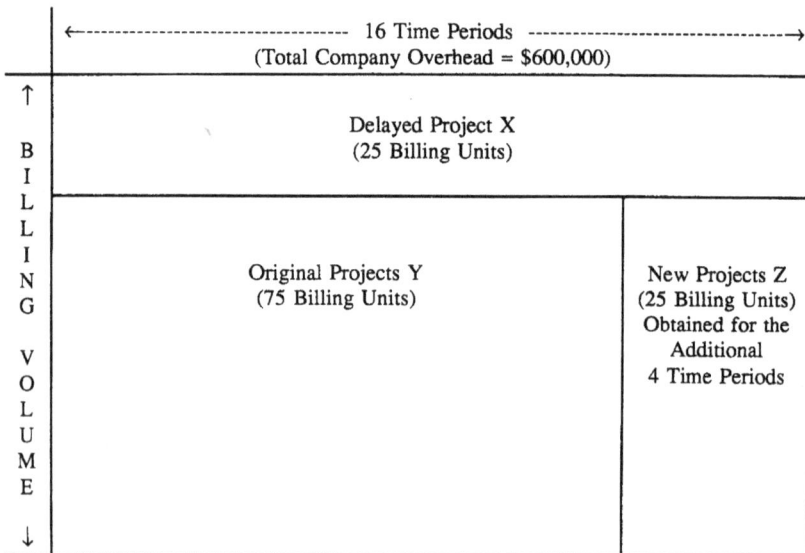

```
    ←----------------------------- 16 Time Periods -----------------------------→
                      (Total Company Overhead = $600,000)

  ↑   ┌──────────────────────────────────────────────────────────────────────┐
      │                          Delayed Project X                            │
  B   │                          (25 Billing Units)                           │
  I   ├──────────────────────────────────────────────┬───────────────────────┤
  L   │                                              │                        │
  L   │                                              │                        │
  I   │                                              │    New Projects Z      │
  N   │            Original Projects Y               │   (25 Billing Units)   │
  G   │             (75 Billing Units)               │    Obtained for the    │
      │                                              │       Additional       │
  V   │                                              │     4 Time Periods     │
  O   │                                              │                        │
  L   │                                              │                        │
  U   │                                              │                        │
  M   │                                              │                        │
  E   │                                              │                        │
  ↓   └──────────────────────────────────────────────┴───────────────────────┘
```

Figure 3-3   Delayed time periods for company overhead and revenue

Figure 3-2 illustrates a normal situation, that is, a situation without delayed projects. In this case, the Eichleay Formula would allocate $112,500 of the company's $450,000 general overhead to Project X and $337,500 to all other Projects Y.

Figure 3-3 illustrates what happens if Project X is delayed by four time periods, that is, 33.33% of its planned time periods. The billing units of Project X (excluding claim billings) remain the same, but the billing units of all other projects have increased by 33.33% for new Projects Z. Evidently, because of its continued involvement with Project X, the company has lost 8.33 new billing units. In this case, the Eichleay Formula would assign $120,000 of the company's $600,000 general overhead to Project X, which is an increase of $7,500, or 6.67%, over the amount allocated for its normal construction period, although the delay has demanded 33.33% more time to construct the project. Even more peculiar, because of the delay of Project X, the Eichleay Formula would also *increase* the allocated overhead for Projects Y in the normal situation by $22,500 (to $360,000). It gets stranger: The Eichleay Formula would allocate an overhead of $120,000 to the new Projects Z, which have no delay period; this overhead is *identical* to that assigned to the *delayed* Project X, even though Project X took four times as long to construct as Projects Z — simply because Project X and Projects Z have *identical* billing units.

It may have been more equitable if the formula had allocated an additional overhead to Project X equal to its additional time period to construct, that is, 33.33% of $112,500, or $37,500. In such a case, Projects Y would not have received an increase in the original overhead allocation, and new Projects Z would have received a more appropriate allocation of $112,500 for their 25 billing units, which is the same overhead that the Eichleay Formula had assigned to Project X and its 25 billing units for the normal construction period (see Figure 3-2).

The flaws in the Eichleay Formula are obvious: In this method, overhead is strictly billing related, without adjustments for time, project size, labor content, or inefficiencies as a result of delays. On the other hand, project overhead decreases (per sales dollar) with larger projects and increases with more labor-intensive projects (see Figure 3-1). Furthermore, complex and troublesome projects take more project overhead than do simple and smooth-running projects. The Eichleay Formula disregards all of these considerations.

We have already seen earlier that the determination of project overhead is fairly simple, both during tender and during construction periods. Therefore, I would like to propose another method that makes use of project overhead for allocating general overhead to various projects. I shall call this method the Thormann Formula.

THE THORMANN FORMULA

Simply stated: $I \times B / J = H$

Where:

$I =$ The delayed project's project overhead for the delay period.

$B =$ The total company overhead during the delayed project's construction period.

$J =$ All of the company's project overheads during the delayed project's construction period.

$H =$ The amount claimable for unabsorbed company overhead.

The Thormann Formula is based on the assumption that there is a correlation between project overhead and general overhead*. This assumption makes sense when you consider the following.

General overhead is also involved with projects, namely, to establish the company's revenue, which consists of projects. Even during the marketing stage, larger projects take less time per sales dollar than do smaller projects, and more labor-intensive projects take more time than do less labor-intensive projects, and so forth. If the tendering process were 100% successful, then the time and money spent on marketing, per project, could simply be computed and charged to each project. However, not all tenders are successful, and the time and money spent on unsuccessful tenders must also be absorbed. For these reasons, a good case can be made that general overhead should attach itself in a prorated fashion to project overhead.

Thus, projects with larger project overheads per sales dollar are allocated larger portions of general overhead. As we have seen earlier, project overhead increases with labor-intensive, complex, and/or troublesome projects, which is the case with delayed projects. Therefore, delayed projects should also receive higher allocations of general overhead than should projects that were not delayed and ran smoothly. The Thormann Formula accomplishes this: because project overhead has inherently increased on a delayed project, the assigned general overhead is also increased, and, since project overhead accounting is fairly simple, the formula makes both overhead validations for delay claims easy.

---

* Company-related overhead (or company overhead, for short) is the overhead that exists without any projects. It is mainly marketing related and is also known as general overhead.

EXAMPLES OF CALCULATING DELAY OVERHEADS

Let us take the situation illustrated in Figures 3-2 and 3-3 and use the following budget and cost data, assuming that the costs for Projects Y and Z were within budget:

Budget and Cost Data for Projects X, Y, and Z
(Including Change Orders)

Project X Budget Data	
Revenue	$1,993,000.00
Materials and Subcontracts	1,266,000.00
Direct Labor	422,000.00
Project Overhead	112,500.00
General Overhead	112,500.00
Interest	20,000.00
Shareholder Profit and Risk Allowance	60,000.00
Projects Y Budget Data	
Revenue	$5,979,000.00
Project Overhead	337,500.00
General Overhead	337,500.00
Projects Z Budget Data	
Revenue	$1,993,000.00
Project Overhead	112,500.00
General Overhead	112,500.00
Project X Actual Costs	
Materials and Subcontracts	$1,266,000.00
Direct Labor (increased for productivity loss)	480,000.00
Project Overhead (increased for time loss)	150,000.00
Interest (increased for higher costs and more time)	28,000.00

Table 3-2 Company budget for projects

To calculate the Project X general overhead for the delay period by the Eichleay Formula:

Step 1: $1,993K \times 600K / 9,965K = 120K$
Step 2: $120K / 16 = 7.5K$ (using time periods for calendar days)
Step 3: $7.5K \times 4 = \$30,000$ claimable overhead,

and, by the Thormann Formula:

1 Step: $37.5K \times 600K / 600K = \$37,500$ claimable overhead.

NB: since the Eichleay Formula (in Step 1) allocates only $7,500 over the budgeted general overhead, a claim recipient may disallow more.

INTEREST CHARGES

Tables 3-3 and 3-4, as shown below, illustrate how interest should be calculated for delay claims. The first table shows the calculations that should have been made during the tender period, and the second table shows what might have occurred during the construction period. The dollar amounts are rounded for ease of illustration. For the same reason, uniform monthly costs and interest charges of 1% per month were used.

Transaction Estimate for the Tender

Month	Costs Incurred	Contract Billings	Contract Payments	Balance Payable		Estimate Interest	
1	10,000	10,000	0	10,000		0	
2	10,000	10,000	9,000	11,000		100	
3	10,000	10,000	9,000	12,000		110	
4	10,000	10,000	9,000	13,000		120	
5	10,000	10,000	9,000	14,000		130	
6	10,000	10,000	9,000	15,000		140	
7	10,000	10,000	9,000	16,000		150	
8	10,000	10,000	9,000	17,000		160	
9	10,000	10,000	9,000	18,000		170	
10	10,000	10,000	9,000	19,000		180	
11	0	0	9,000	10,000		190	
12	0	0	10,000	0		100	
13	0	0	0	0		0	
Totals	100,000	100,000	100,000	0		1,550	

Table 3-3 Typical interest charges in estimates

Actual Transactions Incurred

Month	Costs Incurred	Contract Billings	Contract Payments	Balance Payable	Actual Interest	Estimate Interest	Claim Interest
1	12,000	10,000	0	12,000	0	0	0
2	12,000	10,000	9,000	15,000	120	100	20
3	12,000	10,000	9,000	18,000	150	110	40
4	12,000	10,000	9,000	21,000	180	120	60
5	12,000	10,000	9,000	24,000	210	130	80
6	12,000	10,000	9,000	27,000	240	140	100
7	12,000	10,000	9,000	30,000	270	150	120
8	12,000	10,000	9,000	33,000	300	160	140
9	12,000	10,000	9,000	36,000	330	170	160
10	12,000	10,000	9,000	39,000	360	180	180
11	0	0	9,000	30,000	390	190	200
12	0	0	10,000	20,000	300	100	200
13	0	0	0	20,000	200	0	200
Totals	120,000	100,000	100,000	20,000	3,050	1,550	1,500

Table 3-4 Actual interest charges

The Costs Incurred column includes overheads and return on investment. The interest charges are made on the previous month's balance.* In Table 3-3, the Balance Payable column normally shows only the accumulated holdbacks and the outstanding current payment. The Balance Payable column in Table 3-4 also includes the accumulated costs that were not included in the contract billings, because they are not yet part of the contract. These costs are incurred because of productivity and overhead losses for disruptions and delays, and because of the overtime that had to be worked to maintain the construction schedule. Naturally, this monthly cost can vary substantially, but it is averaged here for ease of following the required calculations.

This format of calculating interest charges can also be used for late payments, but is most useful for establishing the interest charges for claims. The main reason for this is because owners and the courts have a habit of allowing interest charges only on the claim amount *after* the claim has been submitted, if they allow them at all, and forget that the claim costs were incurred much earlier — sometimes right from the beginning of construction, as shown in Table 3-4. In this case, $1,500 is the appropriate amount of interest to be included in the claim.

However, care must be taken that interest charges do not duplicate return on investment (ROI) when the contractor's financial operations depend wholly or in part on ownership capital; in this case, interest and ROI can be mutually exclusive. The last section provides this analysis.

PROFIT AND RISK ALLOWANCES

A few words should be devoted to profit, because it is probably the most widely misunderstood type of charge. Many people think of profit as some kind of bonus or gratuity (sometimes unjustified) that companies want to stash away for their shareholders, and very seldom is it considered a necessary charge to maintain a business.

We all know that every company must be financed, for various purposes, to stay in business. For a construction company, these purposes include financing the company's marketing efforts before it obtains the projects to be built, and then financing all projects' overheads, material purchases, payroll costs, and holdbacks before getting paid by its customers. If the company had no equity investments, all financing would have to come from the likes of banks (assuming they were willing to put it up) and the financing costs would be just that: *costs* to the company that would seldom be questioned. It is hard to understand,

---

* Banks can add interest charges at various times (e.g., daily, monthly, quarterly) but monthly charges are the most common.

therefore, that shareholders, who finance the company with equity capital and expect a return on their investment, should be considered lesser investors than banks, especially when these shareholders secure the bank loans.

Profit is made up of the shareholders' return on their capital investment and risk allowances. Risk allowances are tied to profit because profit (past, present, and future) diminishes when risks materialize. A common mistake is to believe that risks can be paid for out of allowed overheads. When a risk materializes, it can no more be paid for out of overhead allowances than it can be out of direct material or direct labor allowances. Each allowance is a cost that is already earmarked for expenditures in its own right.

There are usually two types of risk allowance: the known-risk allowance (KRA), which is for risks likely to materialize that can be individually assessed, and the unknown-risk allowance (URA), which is for unforseen risks and is usually matched to ROI.

An allowance for unknown risks is usually included with the profit markup, along with ROI, but the various allowances for known risks are often included with other parts of the estimate.

For example, an allowance for extreme weather conditions might be included in the allowance for direct labor, and an allowance for uncollectible shipping damages might be included with the freight for materials. Nevertheless, a shortage in the allowances for known risks must still be paid for out of profits. Because the KRA is made up of individual risk items that are separately assessed, it can vary substantially from job to job (see Figure 3-4). Table 3-5 shows a few examples of the calculations* made for some of the risks assessed for the KRA.

Profit lost because of lost revenue (see Figure 3-3) must also be included as a delay-claim item, because the revenue lost due to the delayed project should have been there to generate this profit. For a calculation example of claimable profit, refer to the data in Table 3-2, for which each billing unit of Figure 3-3 was converted to $79,720.00. Also, Figure 3-3 illustrated that 8.33 billing units were lost as a result of the extended time to complete Project X, which is the equivalent of $664,333.33, and since this lost revenue does not need an allowance for unknown risks, it is reasonable to drop the profit percentage on revenue from 2% to 1%; therefore, the claimable profit comes to $6,643.33. However, as explained in the next section, an additional adjustment to profit may be required when interest charges are claimed.

---

* The percentages shown in Figure 3-4 are based on revenue; if the percentages used in Table 3-5 are based on other cost allowances such as direct labor or materials, a notation should be added to this effect.

Profit Components

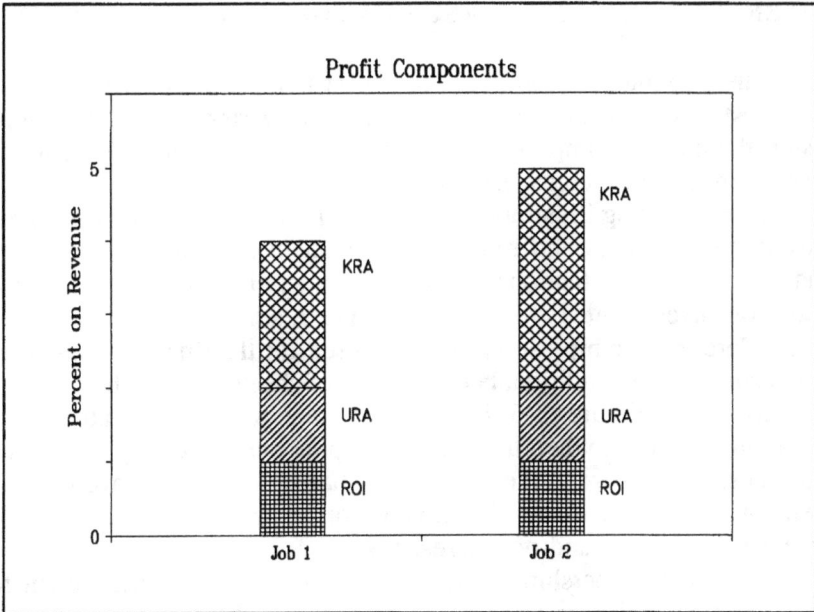

Figure 3-4

Known-Risk Allowances

Check	Type of Risk	Percent
✓	Extreme Weather Conditions Force Majeure Provisions Possible Delays Foreign Territory Currency Exchange	1.3
✓	Remote Location Accessibility Availability of Labor Force	0.8
✓	Available Transportation Poor Designs Inexperience Cooperation Between Various Parties Payment Habits & Solvency of Owner Required Insurances Indirect Damages	0.4
	Total	2.5

Table 3-5

INTEREST CHARGES VERSUS RETURN ON INVESTMENT

The previous section made the point that profit consists of return on investment and allowances for risks. Since risk allowances would distort the interest comparison that is made in this section, only return on investment is being considered here.

The following illustrations are offered for a better understanding of why interest charges and return on investment are mutually exclusive, and of the various situations that would attract either interest charges or return on investment. Figure 3-5 illustrates a contractor's ideal balance sheet. Here, ownership capital is *solely* used for the financial operations of the company. In addition, because balance sheets usually show month-end positions, Figure 3-6 is included here as an illustration of a midmonth position, when the working capital shrinks to pay off current liabilities. The reward to the company's shareholders, who have invested their capital, is the ROI. No money was borrowed to subsidize the financial operations and no interest was paid.

However, ownership capital may, occasionally, be insufficient for all of the financial operations of the company, and the company may be forced to borrow capital to cover this deficiency, as shown in Figure 3-7. This requirement usually occurs in small start-up companies, where individual owners may be in a position to supply personal guarantees to financial institutions supplying the additional capital to operate the company. Since this borrowed capital is used only for the financial operations of the company, that is, the needed working capital, the interest charges on it are chargeable to the company's various projects, and the return on investment generated by the various projects can be reduced proportionately.

Sometimes, a company borrows capital for investments such as land and building purchases. The ownership capital may be sufficient for the financial operations of the company, but not for any additional investments (see Figure 3-8). The interest paid in this case is not directly chargeable to the company's various projects but becomes part of its general overhead, which is proportionately allocated to these projects, and the full return on investment from the various projects is applicable.

However, if the borrowed capital is not only for the company's investments but also for its financial operations (see Figure 3-9), part of the interest would be chargeable directly to the company's various projects, and the other part to its general overhead. Here again, return on investment from the various projects can be reduced proportionately.

Conversely, return on investment from the various projects must be carefully assessed in situations when the company's ownership capital is in excess of the required working capital (see Figure 3-10) and the

Contractors' Balance Sheets

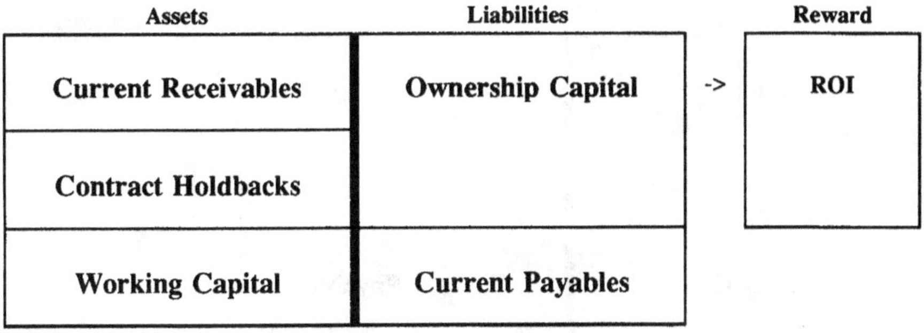

Assets	Liabilities	Reward
Current Receivables	Ownership Capital ->	ROI
Contract Holdbacks		
Working Capital	Current Payables	

Figure 3-5   Fully operational solely with ownership capital (month-end position)

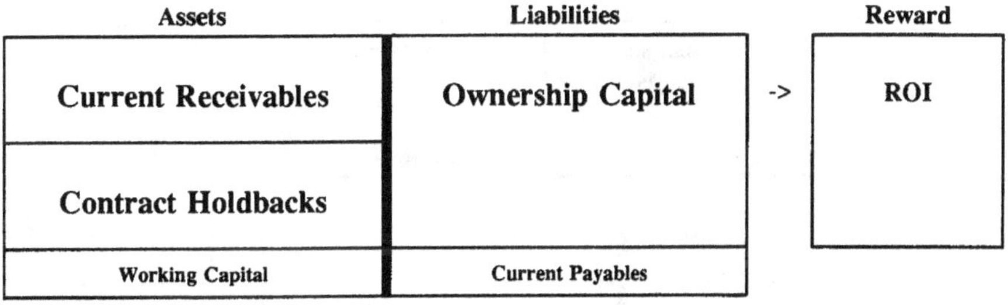

Assets	Liabilities	Reward
Current Receivables	Ownership Capital ->	ROI
Contract Holdbacks		
Working Capital	Current Payables	

Figure 3-6   Fully operational solely with ownership capital (midmonth position)

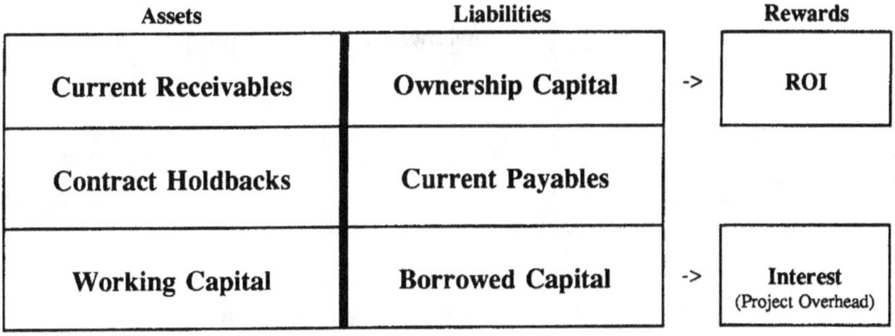

Assets	Liabilities	Rewards
Current Receivables	Ownership Capital ->	ROI
Contract Holdbacks	Current Payables	
Working Capital	Borrowed Capital ->	Interest (Project Overhead)

Figure 3-7   Fully operational only with added borrowed capital

surplus is invested in, for example, land and buildings. In this case, the company is penalized by the higher return expectations of its shareholders vis-à-vis the lower interest charges that could be obtained from financial institutions. It is fair, therefore, that appropriate adjustments are made to the return on investment by the various projects.

Contractors' Balance Sheets

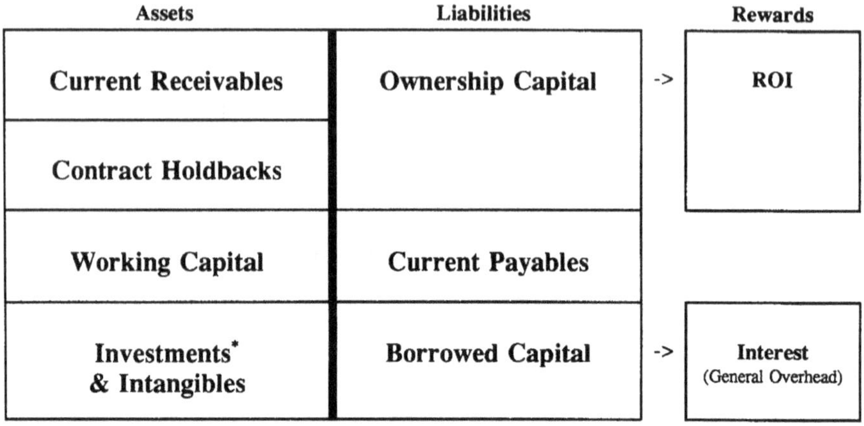

Assets	Liabilities	Rewards
Current Receivables	Ownership Capital	-> ROI
Contract Holdbacks		
Working Capital	Current Payables	
Investments* & Intangibles	Borrowed Capital	-> Interest (General Overhead)

Figure 3-8   Fully operational solely with ownership capital — borrowed capital is required only for investments*

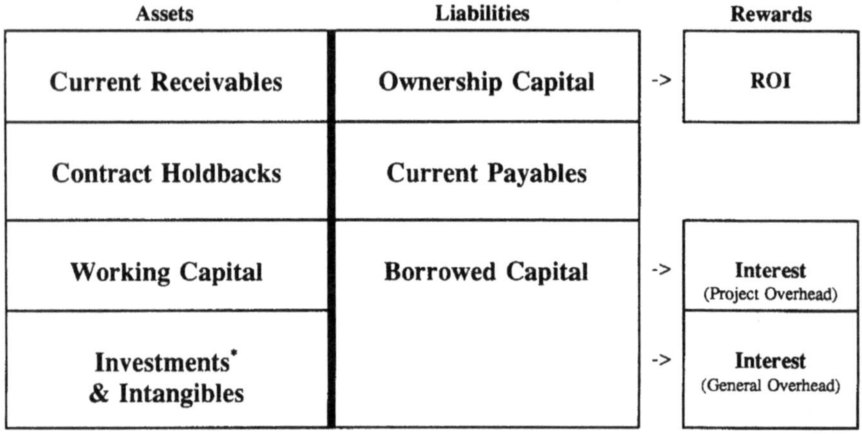

Assets	Liabilities	Rewards
Current Receivables	Ownership Capital	-> ROI
Contract Holdbacks	Current Payables	
Working Capital	Borrowed Capital	-> Interest (Project Overhead)
Investments* & Intangibles		-> Interest (General Overhead)

Figure 3-9   Fully operational only with borrowed capital — which is also required for investments*

A normal ROI is the return that is expected by the company's shareholders on their capital invested. If the company has ownership capital of $250,000 for a yearly revenue of $2,500,000 and the shareholders expect a return of 10% per annum on their investment, then the normal ROI would represent 1% of the company's yearly revenue.

---

* Investments could include land, buildings, furnishings, equipment, tools, stock, and so on.

Contractors' Balance Sheets

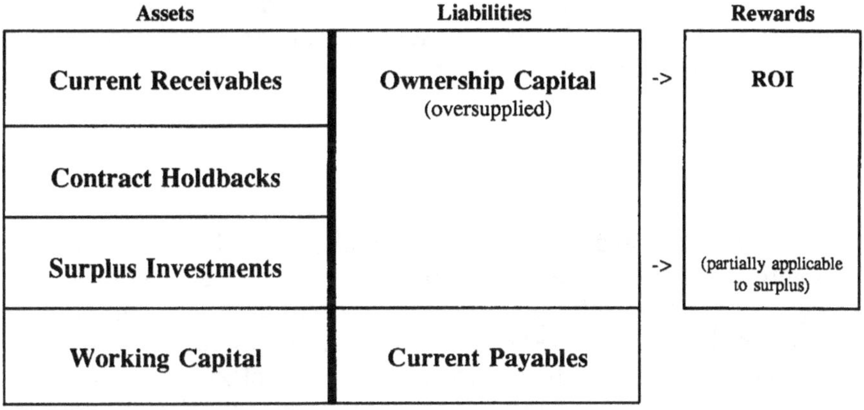

Assets	Liabilities		Rewards
Current Receivables	**Ownership Capital** (oversupplied)	->	**ROI**
Contract Holdbacks			
Surplus Investments		->	(partially applicable to surplus)
Working Capital	**Current Payables**		

Figure 3-10   Fully operational with ownership capital in surplus

Since interest is directly chargeable to projects only in cases when borrowed capital is required for the financial operations of a company, when the return on investment can be reduced proportionately, it is useful to apply a formula to the normal return on investment (NR) to arrive at the expected return on investment (ER) from the various projects:

$$ER = (1 - (BC - IN) / (BC - IN + OC)) \times NR$$

For example, if a company has $200,000 of shareholders' ownership capital (OC) and acquires $200,000 borrowed capital (BC) of which $50,000 is required to subsidize working capital and $150,000 is required for investments (IN) in land and buildings, the amount of ER from its projects is arrived at by multiplying the NR by 0.8. In this example, the normal return on investment is reduced by 20%, although it still provides the 10% per annum return expected by the shareholders on their capital.

The interest charges, in this example, are prorated so that 25% is charged directly to the company's projects and 75% to the company's general overhead — providing, of course, that the interest charges on the entire $200,000 of borrowed capital are identical; otherwise, a further adjustment is necessary.

It is evident from the above illustrations and examples that interest charges and return on investment are mutually exclusive. Therefore, if the full profit is allowed on a project, interest should not be charged, and, if interest is chargeable, the profit should be reduced accordingly.

However, because these illustrations apply only to the normal course of business, when material and labor is supplied and billed for in one month and paid for in the next month (except for holdbacks), interest is still chargeable if payments are delayed, such as for disputed claims, which are not settled for several months or several years. In this case, additional capital must be borrowed to cover the abnormal, outstanding account receivable.

CHAPTER 3 REVIEW

Project overheads are easily identifiable, even when the company's accounting system has neglected to do so, but general overhead must always be subjectively allocated to various projects. The easiest allocation of general overhead to projects is by using a uniform percentage, but this method is inequitable. Complexities and required attention vary substantially from project to project, and delayed projects are usually at the higher end of the scale and should receive a higher allocation (per revenue dollar) than should projects that finish on time.

The Eichleay Formula attempts to allocate general overhead but is, as we have seen, not very successful – for the simple reason that it is using the uniform markup method. Tying general overhead to project overheads is much more equitable and can be accomplished by using the Thormann Formula. Project overheads vary considerably from project to project (per revenue dollar) and the very disposition of a delayed project requires higher project overheads. It makes sense, therefore, to allocate higher general overhead to delayed projects as well.

Interest charges and returns on investment, too, must be properly allocated to make a convincing case for their claimability. Recipients of claims are fairly sensitive when it comes to these charges – and so are the courts. Claimants are well advised to present them fairly and reasonably, especially when interest charges are involved. Keep in mind, however, that a fair profit, as a return on the shareholders' investment, should be treated with the same respect as interest payments on bank loans. Furthermore, when future profit is lost due to lost revenue because of a delayed project, it is very much a claimable item.

4

Preparing Information
for
Construction Delay Claims

GENERAL COMMENTS

The information that must be gathered and prepared for a construction delay claim is for the purpose of backing it up. One party accuses another of being responsible for schedule disruptions and delays, and must provide evidence in support thereof. Conversely, the recipient of the claim usually produces evidence to refute it. Such evidence, by either party, requires what is known as *documentation*.

The party advancing the claim will also want to back up its quantifications by producing accurate records of the major costs involved, namely, how productivity losses and the additional overheads were calculated. Since such claims are usually in the six and seven figures, it is only natural that these costs will be scrutinized critically by the recipient.

Another issue usually addressed by both parties, from opposite viewpoints, of course, is the question of entitlement. The claimant will try to find support in the contract documents* to justify his claim. The recipient, on the other hand, may refer to some general conditions of the contract to point out that there is no entitlement, which may not be too hard, because contract conditions are usually fairly strict and restrictive for both parties.

The claim, in the final analysis, may hinge on how well the claimant is prepared and how authentic his documentation is. The preparation should, ideally, take place throughout the construction period. Whenever preparations are done *after* construction is completed, the task is not only harder but all too often frustrating, because only insufficient or inadequate support for the claim may be available at this time, even though all parties know exactly what has happened.

Many claimants are often sadly unprepared for claims, and whatever little or lacking documentation can be found, for whatever it is worth, must still be established at this late date (when construction is completed). This task usually requires the help of an experienced, outside claims expert.

The following sections offer the details necessary for preparing such information.

---

* Quite often, various authorities, such as publications by claims experts and settled court cases, may also be quoted.

DOCUMENTATION DURING CONSTRUCTION

A construction project generates an abundance of documents, and it pays to keep them well organized. Good claims management generates even more documents, which should be filed separately from routine documents. The success of a company is usually enhanced by well-organized and conclusive documentation. Whenever I observed remarkable success by project management, it was always accompanied by a well-organized filing system and conclusive documentation. Those project managers who achieved a degree of success without conclusive documentation were usually lucky.

The following documentation is helpful, and sometimes essential, to good claims management and claims preparations:

- **Site memos** relating to construction delays and schedule deviations; these are notifications of a sort but are less formal than letters.
- **Site pictures** (with dates) showing construction progress and signed on the back by a supervisor, with a note of explanation, although the pictures should be carefully chosen to be self-explanatory, that is, relative to the construction schedule.
- **Letters of notification** of delays and impacts. These are more formal than are mere site memos and provide more information about impacts or potential impacts. Letters of notification may also request contract change orders for schedule extensions, which may not be heeded but are there for the record. Often, contract conditions will specify the forms and time frames of such notifications, and project managers are well advised to follow them.
- **Requests for clarifications** (RFCs) may not be related to project delays, strictly speaking, but are nevertheless required to establish the disruptive effect of a poorly designed project. If too many of these requests were necessary, they provide good evidence for delay causes and reason for recompensation.
- **Requests for change orders** to the contract, including requests for schedule extensions and payments for impacts.
- **Notifications of schedule changes** and their responses. Many players are involved when it comes to schedules, and everyone deserves to be notified of schedule changes even if he is not involved in the immediate time frame. The responses to such notifications may provide needed evidence in future claims.
- **Minutes of meetings**, with a separate list of references to notations regarding schedule disruptions, delays, impacts, and extensions. Since such minutes are (or should be) circulated to

all parties involved in the construction project, they are highly regarded as good notifications of delays and require responses for discrepancies.

- **A log record of purchase orders** with requests for shop drawings, including the shop drawing receipt from the supplier, the submission and return for approval, and the return to the supplier. This documentation will often provide the proof of who is, or was, delaying the process, which is invaluable in case of future claims.
- **A log record of requests for quotes on contract changes,** and the submission, acceptance, or rejection of these requests. Again, if delays in the process occur, this record will establish the culprit.
- **Tracking the construction schedule,** with actual manpower loadings. It goes without saying that the as-planned start and finish times of various tasks should be compared with the as-built times; it is also useful to record the manpower loadings to establish the impact of delays.
- **Progressive Construction Completion Reports** are treated in detail under Task-per-Area Assessment reports in Chapter 2; they are included here for their documentation value.
- **Weekly Field Progress Reports** are treated in detail in Chapter 2; they are included here for their documentation value.
- **Disruption and Delay Impact-Assessment Reports** are treated in detail later in this chapter; they are included here for their documentation value.
- **An accurate record of project overhead.** Project overheads have already been described in Chapter 3, but it is important that they be properly documented. Many project overheads are time dependent, which should be recorded in the documentation.
- **The expediting record of material shipments** and deliveries, including the agreed-on shipping dates in purchase orders.
- **The master material list,** which includes all estimated and purchased materials and their purchase orders. This list is not only important to project management to establish estimate deviations — and to make appropriate planning adjustments — but also to the claim preparations discussed in Chapter 5.
- **The job-site diaries** of field supervisors, which provide good evidence if kept up regularly and impartially.
- **Payroll records,** which are sometimes needed to provide evidence.

DOCUMENTATION AFTER CONSTRUCTION

It is often difficult, if not impossible, to find or to establish good claim documentation after a project is completed. All too often, construction companies, especially trade contractors, are not properly prepared to issue claims — even their accounting systems are poorly designed for claims — which makes it almost impossible to back up their assertions. Nevertheless, *some* documentation is usually required, and the staff assigned to help the claims expert[*] must spend many hours trying to find or establish it — sometimes to no avail, because the claim recipient may not accept its authenticity.

After construction is completed, the following process is usually required to assemble whatever documentation may still be available:

- **Site memos,** if they exist, must be reread to segregate those pertinent to the claim.
- **Site pictures** may not exist at all if delays were not anticipated; check with other contractors who were on site.
- **Letters of notification,** if they exist, may be buried in the general correspondence files; dig them out.
- **Requests for clarification** are seldom relevant to delay claims, except to prove the number of disruptions caused by poor design.
- **Requests for change orders** are usually required to establish the magnitude of delay causes, and should be referenced.
- **Notifications of schedule changes,** if they exist, are important to delay claims; dig them out of the files.
- **Minutes of meetings** may contain references to somebody's delays; they must be reread to establish a list of such references.
- **A log record of purchase orders** with requests for shop drawings can often be established from suppliers' data if the contractor did not keep one and if late shop-drawing approval is a delay cause.
- **A log record of requests for quotes on contract changes** can also be established by obtaining information from others; this log record may be important to establish holdups of construction areas that occurred while the contractor prepared quotes and

---

[*] Larger companies may have a person on staff who is familiar with claims and would be assigned as the claims manager whenever a claim must be prepared, in which case good documentation was probably assembled during construction, and an outside claims expert would merely write an opinion. If no staff person familiar with claims is available, the claims expert must take a much more active role in the claim-preparation process.

waited for approvals — even if the quotes were rejected.

- **Tracking the construction schedule**, with manpower loadings for the various tasks, may be impossible to establish (at least accurately) unless this was done during construction; however, a record of overall manpower loading can still be prepared.
- **Progressive Construction Completion Reports**, or Task-per-Area Assessments, even if they could be established after construction is completed, which I doubt, are probably meaningless because they would be too subjective; therefore, forget about them.
- **Weekly Field Progress Reports** provide the same dilemma; they should be completed during construction.
- **Disruption and Delay Impact-Assessment Reports**, which are treated in the next section, can probably still be prepared at this point — provided good data are still available — however, their accuracy is much more reliable if prepared during construction.
- **An accurate record of site overheads** can usually be faithfully established after construction is completed, but at a much higher cost and investment of time.
- **The expediting record of material shipments and deliveries** is hard to establish after construction is complete unless kept in the field supervisors' diaries, in which case it should be listed separately for easy reference.
- **The master material list**, if it does not exist, may be extremely hard to establish in its entirety; however, the claim recipient usually requires at least a comparison between estimated and purchased materials. Again, this can be time consuming after construction.
- **The job-site diaries** of field supervisors, if they do not exist, are impossible to establish after construction (with credibility); at best, the contractor could obtain notarized statements from supervisors.
- **Payroll records** should have been kept with segregated information on each project; if this is not the case, records may have to be established from time sheets. Besides being useful to the claimant in establishing impacts, the claim recipient may request these records to verify the claimant's assertions.

Needless to say, the work required to establish all of this documentation after completed construction is enormous. A claimant may wish (usually too late) that he had never started. The work may only pay for itself if the claim is very large. Therefore, it would be best to get an estimate from a claims expert *before* plunging into this process, which will only get worse as the claim proceeds to discoveries.[*]

A word of caution is in order. Quite a bit of the documentation exists somewhere (perhaps with another party); it is just time consuming to establish and assemble *after* construction. However, if there is no documentation for some element of the claim that requires a reference, do not, under any circumstances, create it at this stage. The best alternative would be to obtain a sworn statement from someone who has direct knowledge of the circumstances that prevailed.

ASSESSING AND ALLOCATING PRODUCTION LOSSES

The best time to assess and allocate production losses is when they occur. This reason is why field supervisors should be encouraged to fill out a report (such as that shown in Figure 4-1) every time the work experiences disruptions and delays.

Suppose a six-man crew has been mobilized to install fixtures in an area, and it is discovered that the fixtures are defective. The field supervisor will have to demobilize the crew and remobilize it in another area for a different task. The best time to assess the time lost and to whom to allocate it is right then. The form (Figure 4-1) is filled out by the supervisor, who gives a detailed explanation and lost-time calculation, signs and dates it, and then submits it to the project manager, who processes it further.

Some of these disruptions and delays may be minor, and the project manager may just issue a warning to the party responsible; however, if these disruptions and delays have a more serious impact, or if the minor disruptions and delays repeat too often and cause a serious cumulative impact, the project manager may issue a claim for compensation (after giving the appropriate notice). Completing an immediate impact assessment whenever disruptions and delays occur will enable all parties involved to take timely remedial actions and, thereby, mitigate the losses.

Another advantage of this reporting method is that field supervisors are requested to assess any and all disruption and delay impacts without selection. In other words, if the company itself caused the problem, it would be similarly assessed. There are seldom, if ever, time losses that occur with only one culprit responsible, and when these assessments are reported fairly, the contractor's case for a delay claim will be much stronger, because the final summary will also show those losses that the contractor was willing to absorb or was able to pass on to others.

---

[*] Discoveries are required before proceeding to trial, and, before proceeding to discoveries, a complete list of documents must be made available by each party. If this list is incomplete, more information is usually requested.

ABC CONSTRUCTION CO.

LOST-TIME IMPACT ASSESSMENT

LOST TIME DUE TO:	ASSESSED* TO:							LOST-TIME CALCULATIONS:
	O	E	C	T	A	S	I	
LATE DELIVERIES SCHEDULED AREAS NOT READY AREAS ON HOLD CHANGE-ORDER DISRUPTIONS ACCIDENTS OR ACTS OF GOD								
BAD DESIGNS ABNORMAL OBSTRUCTIONS ABNORMAL WEATHER CONDITIONS ABNORMAL SITE CONDITIONS INADEQUATE INSTRUCTIONS								
IMPROPER PLANNING / LAY-OUT POOR WORKING DRAWINGS POOR SHOP DRAWINGS IMPROPER SCHEDULING POOR COMMUNICATIONS								
POOR PAPER FLOW INTERFERENCE FROM TRADES INTERFERENCE FROM OWNERS INCOMPETENT SUPERVISION INADEQUATE SUPPORT STAFF								
INCOMPETENT WORKFORCE TRAINING / ORIENTATION / SAFETY LUNCH & REST AREAS TOO FAR TOO MUCH NONPRODUCTIVE TIME BAD MORALE								
INSUFFICIENT TOOLS & EQUIPMENT INAPPROPRIATE TOOLS & EQUIPMENT POOR MATERIAL DISTRIBUTION STORAGE AREAS TOO FAR ABNORMAL REPAIRS								
MOBILIZE / DEMOBILIZE / REMOBILIZE OTHER (specify):								

DETAILED EXPLANATION:

_____

_____

_____

HOURS (from above): _____ @ RATE: _____ = TOTAL COST = _____

PROJECT:_____SUPERVISOR:_____DATE:_____

* O = Owner, E = Engineer, C = Contractor, T = Trade (other), A = Act of God, S = Supplier, I = Internal (own)

Figure 4-1

Another advantage of using this time-loss assessment method is that it enables the company to assess its labor estimates more accurately. All too often, companies assume that a few known disruptions that occurred during construction were responsible for the entire labor overrun, which may or may not be the case.

Finally, an immediate assessment is much more likely to receive fair treatment from the claim recipient than another more subjective assessment that is sometimes established at a much later date. The form shown in Figure 4-1 is useful for this purpose. It is generic and lists only a few of the situations that can cause disruptions and delays. Each company should design its own specialized form. The list is a reminder of examples for supervisors and creates an awareness of various causes. Spend some time with your supervisors on what needs to be recorded.

ALLOCATING PROJECT-RELATED OVERHEAD

Each company's accounting system should provide for segregated project-related overheads. This method not only helps in the preparation of delay claims, but it also helps to put the company in a better competitive position by avoiding the use of averages that may either be too high or too low for any particular project (see Chapter 3). The following is a typical list of project-overhead items that may be affected by construction delays, because their costs are time related:

- **Site Equipment and Facilities**
  - Office trailers (furnished)
  - Office equipment (computers, fax machines, copiers, etc.)
  - Material storage facilities (including off-site)
  - Tool storage facilities
  - Lunch and change facilities
  - Toilet and wash facilities
  - Tools and construction equipment
- **Site Utilities**
  - Telephone-line charges
  - Electrical power charges
  - Heating fuel
  - Water and sanitation
  - Fuel and oil for trucks and construction equipment
- **Project-related Staff**
  - Project manager / superintendent
  - Foremen (general and nonworking)
  - Material coordinator and storage supervisor
  - Engineers / schedulers / draftsmen
  - Clean-up crew

- **Other Project-related Overhead**
  - Insurances
  - Bonding
  - Finance charges
  - Warranties

If any equipment is company owned, the project should nevertheless receive a rental charge for it, since the company is not in business to provide such service free of charge. The charges for equipment that is supplied by rental companies include handling, storing, maintenance, overhead, and profit, and the same applies to construction companies.

It is also important to keep track of the time frames in which equipment and facilities are used, so that the time frames can properly be related to delay periods. This applies to staff as well. The staff should keep track of their time used for various tasks, because some of this time may be related to disruptive elements and delays.

If project overheads are incurred because of schedule accelerations, they may not be time related, and careful notations may be required to identify and justify them. For example, staff may have been required to work overtime, or there may have been additional tools required for increased crews. However, in most cases the additional overhead *is* time related and in direct proportion to the delay. In the latter case, the additional overhead is relatively easy to establish and justify.

CHECKING ON CLAIM ENTITLEMENTS

It is always a good idea to familiarize oneself with the provisions of the contract, especially the provisions with respect to claim causes, such as schedule delays and required remedies. Some contracts are very strict and restrictive in this regard and may require advice from legal counsel (see Chapter 5).

Various clauses from the contract form and the project specifications (such as Instructions to Bidders, General Conditions, and Supplemental Conditions) relating to schedule requirements, delays, and remedies should be extracted and relisted with appropriate comments.

Provisions may have been made for Acts of God and other occurrences out of the contractor's control (sometimes referred to as *force majeure*), and these provisions should be carefully analyzed since they are not always favorable to the contractor. For example, some contracts may not even provide for schedule extensions under these circumstances, and may require the contractor to accelerate the work.

Other provisions may be restrictive when it comes to reimbursements for delays, even when the owner is clearly at fault.

Again, such provisions should be carefully analyzed so that the risks are clearly understood. In some cases, the contract provisions may even provide penalties for delays, regardless of who is at fault. In other words, get the job done on time, no matter what.

Most contracts have specific notification clauses, which should be strictly followed lest potential claims be abrogated. This provision is challenging. For example, suppose you must notify the owner of a delay within five days of its occurrence and you miss this time: is a delay claim still valid? Furthermore, some delays may not have exact demarcation lines: they may occur over extended periods before their impacts are noticed; other delays may have been too small to cause a fuss over, and suddenly their collective magnitude becomes apparent. Notice provisions should, of course, be adhered to as much as possible, unless there are excusable circumstances for having ignored them.

CHAPTER 4 REVIEW

We have seen that preparing documentation during construction is more fruitful than preparing it after construction. Finding good documentation after construction is not only harder but it is often impossible. Assessing and allocating production losses is also best done at the time of occurrence — during construction. If one waits until construction is completed to establish production losses, the process becomes much more subjective and requires different methods than those used during construction.

For allocating project-related overheads to periods of delay or acceleration, a good, segregated accounting system is best, keeping in mind that the use of company-owned tools and construction equipment should be charged to projects as if they were rented.

Finally, claimants as well as claim recipients should be well versed in the contract conditions concerning schedule-delay provisions and entitlements for damage reimbursements. These conditions may also require the opinion of an experienced construction claims lawyer.

5

Quantifying the Damages and Preparing the Claim Submission

GENERAL COMMENTS

Claim submissions can range all the way from a few hundred dollars to a few million dollars. Major claims are relatively rare, but minor claims are more frequent, and they usually take the form of a letter with some attached references. When a claim submission goes beyond a letter, it is usually organized in a coil-bound manual, accompanied by a letter of introduction.

The claim manual consists of a title sheet, an executive summary, a table of contents, and various tab sections under which more details are provided, followed by a quantification summary section; a reference section is provided at the end, organizing the backup documentation and references to authorities under further tabs. Since the quantification of damages can include six or more breakdown items, it is wise to list each of them under a separate tab. It is also wise to introduce the damage quantification in a separate overview section, which can make mention of the mitigation measures that were taken by the claimant. All of these tab sections are listed in the table of contents at the outset. This practice is the simplest way to organize a claim submission, both for convenience of reference and for ease of following its explanations.

Two or three copies of the claim submission manual are usually sent to the recipient, and a similar number is retained by the claimant — the additional copies are submitted for convenience in case experts or legal counsel have to be involved.

Initially, a deadline is seldom demanded for a response — first, because the recipient needs time to digest and investigate the allegations in the claim, and, second, because ultimately the claim may proceed to court, which imposes its own time limits on the process. However, if the recipient of the claim takes an unusual length of time to respond, perhaps to convey his disdain, a follow-up letter imposing a deadline may be in order.

The claim submission is by no means limited to single issues. Outstanding issues over scope-of-work disputes sometimes need to be resolved, and, if this is the case, these issues can be included in the claim manual under separate tabs to avoid cluttering up the main claim issue(s).

The following sections offer details to keep in mind for claim submissions. However, the focus is on technical and organizational rather than on conceptual details, which are covered in other chapters.

THE EXECUTIVE SUMMARY

An executive summary is similar to the introduction of a book. It supplies just enough particulars to make the claim comprehensible without going into the comprehensive details of the sections that follow in the claim submission. However, the executive summary should touch upon all aspects of the claim. It can briefly explain the causes and their effects without actually quoting quantifications, for which a separate summary is provided.

Some executive summaries include references, for example, construction schedules or contract conditions, regarding various obligations and/or entitlements. Other summaries offer subjective conclusions. However, try to refrain from giving opinions and let the facts speak for themselves. If any of the facts are in dispute, suggest a meeting time to resolve them.

Remember, the executive summary can be an ideal step to future relations and negotiations, so use the opportunity to establish these relationships amicably. Should you wish to express criticism, blame circumstances rather than personalities. Furthermore, refrain from ultimatums, but be firm in your assertions. Always keep the business relations angle in mind (see Chapter 7). Do not close any doors, and remember the old expression: you can catch more flies with honey than with vinegar. If you feel like suggesting any settlement options, which is not recommended at this stage, do it on a "without prejudice" basis, but be aware that even this attempt may be a sign of weakness.

DETAILED DESCRIPTIONS OF THE CLAIM CAUSES

The details of the causes are provided in this section of the claim submission. The details should be as comprehensive as possible. List them in chronological order, and refrain from giving opinions; remember, stick to the facts and do not volunteer or omit anything that could embarrass you during future discoveries.

There are causes that result only in delays. Other causes result in work disruptions and delays; these work disruptions can result in further delays because of the lowered productivity. If the delay occurred before an activity started, there is usually no work disruption, although there can be if it took place during mobilization. For these reasons, it is best to separate the causes as much as possible. Experience has shown that it is seldom productive to include too many causes in a general summation; besides, this overkill is usually viewed with distrust. Nevertheless, separating the various causes is often hard, perhaps because they are intermingled and overlapping, and all-inclusive statements may become unavoidable. In such circumstances, provide as much detail as possible.

The details should include a comprehensive cause description, the time of occurrence, the duration of delays, and the nature and type of work disruptions, if any. Some causes may have occurred repeatedly; in this case, describe them repeatedly, with their differing situations, time frames, and so forth. As much as possible, refrain from generalizations: specifics carry more clout.

If several causes occurred simultaneously but are unrelated, make sure that they are separately described; and if they are assessable to different respondents, include them in each claim submission for the record, to acknowledge their existence and disposition. If one or more delay causes are due to the shortcomings of the claimant, they must be fully acknowledged at this point, to avoid accusations of inflating the claim. If the owner's consultants and subconsultants are involved in the claim causes, they are usually treated as the owner's agents; therefore, their causes are included in the claim against the owner.

The temptation during the description of the causes is to drag in the effects; this practice may lead to emotionalism and confuse the issues at this point. Unless their mention here serves a useful purpose, the effects, and their quantifications, should be treated in separate sections. By the same token, do *not* refer to obligations and entitlements at this stage — stick to the causes. Comparisons, to strengthen points, are also better placed in the section on effects. However, some type of references (for example, letters) may be in order in this section, especially if the causes are expected to be discredited. It is always wise to clarify nebulous causes as much as possible.

OVERVIEW OF THE EFFECTS AND THEIR MITIGATION

Under this heading in the claim submission, we move from the causes to the effects and, partially, into quantifications. I say "partially" because the main part of quantifications, the financial part, is detailed in subsequent sections of the claim submission. In general, construction-delay effects include, in addition to revised construction schedules, additional time-related overheads (see Chapter 3), material-price escalations, extended warranties, and labor-related losses such as lost productivity, wage-rate escalations, and employee-benefit increases.

If one of the claim causes was the unavailability of the site to start construction, the effect of this delay is a postponed completion date and/or construction escalations. In this case, because neither mobilization nor construction had started, there would have been no work disruption with its associated effects; therefore, all quantifications are time related. This delay effect can be mitigated (usually at the owner's request to obtain the facility sooner) by accelerating the work once it is underway. However, an acceleration has its own adverse effects (see Chapter 2). Mitigation measures that were taken to reduce the effects of the causes should also be detailed in this section.

Each of the causes listed in the claim submission should be separately addressed in this section to establish all of the resulting effects. If causes occurred simultaneously, they may have common effects that must be relatively apportioned. This separation is important because different parties may be responsible for the various causes. However, care should be taken to avoid double-dipping. Delay causes may have cancelling effects: for example, a project is put on hold and a labor strike occurs at the same time; and some effects may be mutually exclusive: for example, if a project is delayed, the contractor may also delay material shipments (at the risk of price escalations) to avoid site storage, so additional or prolonged site storage should not be listed as another effect. On the other hand, if causes occurred repeatedly, their effects are often cumulative, and it may not serve a useful purpose to separate them, even if it is possible to do so.

Mitigation is intended to reduce the effects of the causes but, as mentioned earlier, it usually has its own secondary effects. For example, if an area is put on hold and the contractor lays off workers who may not return, additional training is required for new workers — even workers who return will have to undergo some retraining or reorientation. Therefore, any adverse effects of mitigation should be listed in this section along with its beneficial effects.

The hardest job may be the apportionment of the effects of delay causes to various responsible parties. However, this assignment is especially critical if the claimant himself had shortcomings that caused delays and/or losses of productivity and overheads. Nevertheless, such effects must also be detailed as honestly as possible, lest the claim submission becomes questionable and controversial.

Nebulous effects should be explained as much as possible. For example, you may wish to demonstrate in a table the effects of a number of delays due to required requests for clarifications (see Table 5-1). Such a table gives a good overview of planned and actual construction times, planned and used-up float times, the number of RFCs affecting each area and the average response times, the delay days of critical construction time (that is, when it was impossible to use up float time), the number of disruptions caused to the crew sizes that were affected, and the resulting loss of manhours of work. Even if it were possible to describe all this detail in words, a table is a better medium to illustrate the detail and effects of delay and disruption causes, and the reconciliation of the various numbers, forcing you to make sure not to overstate your case.

	Plan'd Constr Time (days)	Actual Constr Time (days)	No. of RFCs issued	Avg Resp Time (days)	Float Time used (days)	Crit. Task Delay (days)	# of Task Dis- rupts	Avg Crew Size used	Prod. Time Loss (hrs)
	Analysis of the Disruption Effects of Requests for Clarification (RFCs)								
Area									
A	27	27	2	3	6	0	1	8	32
B	90	97	8	4	24	8	5	10	200
C	90	106	15	3	30	15	11	11	484
D	60	66	12	3	33	3	11	7	308
E	60	73	10	4	24	16	8	6	192
F	30	30	4	5	20	0	4	4	64
Total	357	399	51	3.5	137	42	40	8.45	1280

Table 5-1

Notes:
1) Each area has critical tasks (without float time) and noncritical tasks (with float time). The delay of critical tasks caused the project completion to be extended by 42 construction days (or two calendar months).
2) The delay of critical tasks did not, however, cause additional productivity losses, since its effect was to create float time for these tasks; hence, the only productivity loss was due to task disruptions.
3) The average time lost per crew for each task disruption came to four hours, because of resequencing the workflow, demobilization, remobilization, and reorientation.
4) The planned schedule of 357 construction days was based on an average crew of nine workers and an 8-hour workday. The actual schedule ended up with 399 construction days and an average crew of 8.45 workers. This accomplishment indicates that the company's project management was well in control of potential production losses due to the delays caused by RFCs, although it was unable to further mitigate the production losses due to the disruptions caused by RFCs.

Well-prepared reconciliations of the numbers should lead to logical conclusions. It is important, in this regard, to address all influencing factors — even if they tend to harm the claim. Address the harmful issues with logical explanations and make sure that all numbers make sense and add up properly. If one tries a cover-up of any occurrences harmful to the claim, the entire claim is usually discredited. Always assume that the respondents are at least as knowledgeable as you, if not more so.

Time losses are usually easier to establish than productivity losses. For establishing productivity losses, proven, unrelated examples may have to be cited. It is also useful, on occasion, to quote theories expounded by authors such as C. Northcote Parkinson in his book *Parkinson's Law**. In my experience, no truer words have ever been

---

* First published in Great Britain April 1958 by John Murray. Reportedly, on November 19, 1955, an unsigned article appeared in *The Economist* simply entitled, *Parkinson's Law*; the article began: "It is a commonplace observation that work expands so as to fill the time available for its

spoken. As a manager, I used to shudder when at the end of our estimating time for a tender, the tender period was extended. Invariably, my estimators used the extra time for "fine-tuning" the estimate rather than for starting a new one. Similarly, when a project is retarded because of poor information flow, or other reasons, the amount of work to be done usually fills the prolonged time frame. Needless to say, this stretch-out causes a serious loss of productivity. If it is possible to finish a task in five days and the time available is six days, it usually takes six days.

Parkinson's Law explains only part of the productivity losses due to delays, however; other productivity losses can occur as a result of poor attitudes and loss of morale. A worker's attitude depends a lot on those who facilitate his production: the planners and organizers; the material, tool, and area suppliers; the design and information providers; and so on. The worker starts with a "care" attitude to get the work done on time and within budget, but, as he is more and more disappointed by his facilitators, his attitude changes to "why should I care when nobody else does?" and, with this new attitude, his morale deteriorates as well.

When the morale drops, a loss of productivity is inevitable. The causes and effects of this dilemma should not be underestimated. Morale is usually lowered by some kind of dissatisfaction, like having to rework installations for contract changes, or working in prolonged weather extremes, or being confronted with impossible tasks, or having to work with inadequate tools, just to name a few. Whatever the causes for the loss of productivity, the loss of interest in getting the job done is the worst. When assessing poor performance of a worker or a crew, supervisors often refer to what is known as "the triangle of performance".

KNOWLEDGE

ATTITUDE △ ABILITY

THE TRIANGLE OF PERFORMANCE

Did the workers know the requirements for performing the work? Were workers familiar with the site and construction? Were the required tools available and adequate? Were clean work areas available in time to meet the workflow? Was the information lacking? Were all the materials

---

completion. Thus, an elderly lady of leisure can spend the entire day in writing and dispatching a postcard to her niece at Bognor Regis. An hour will be spent in finding the postcard, another in hunting for spectacles, half an hour in search for the address . . ."

on site in time? Did severe weather conditions hamper the work? Also, were there any causes for loss of interest in the work — either personal or job-related? This question requires careful attention, because it is often the reason for unusually high losses of productivity.

To establish credible productivity losses, one may have to use one or more of the methods suggested in Chapter 2. If two different approaches have similar results, it is hard to argue the point. However, since some of the productivity measurements depend on the estimate, the claimant must make sure that there are no major differences between the purchased materials and the estimate: check on quantity differences and price differences, and check on whether price differences affect installation labor. Since productivity losses are subjective at the best of times, it is wise to include detailed, logical explanations in this section to avoid possible misunderstandings.

VARIOUS DAMAGE QUANTIFICATIONS

The foregoing claim-submission section described the effects of the claim causes, which included the number of lost construction days and lost manhours of production. The next claim-submission section offers the dollar calculations for the various damage quantifications. Each calculation category can have its own tab or can be listed under one tab with a separate heading. Following is a list of categories with a description of the elements that are usually included in these calculations.

Productivity losses due to disruptions and delays: If several causes were involved, list a separate calculation for each, with the number of manhours lost multiplied by the hourly rate. The hourly rate should include the employee benefits and labor burden but no other overhead items or profit, which are calculated under separate headings.

Productivity losses due to accelerations: The same comment applies as for productivity losses due to disruptions and delays. In addition, overtime or shift premiums may have to be added. Remember, working overtime or late-night shifts can produce its own productivity loss and the calculations should reflect this loss.

Additional project overhead due to disruptions and delays: These items are usually time related, such as extended time for tool and trailer rentals, extended time for bonds and insurances, extended finance periods, and so on; they can also be production related, such as more effort required by support staff because of disruptions to the work that took additional administration but not necessarily additional construction time.

Additional project overhead due to accelerations: This category analyses additional requirements for tools and construction equipment due to increased crews, additional field supervision, overtime by support staff, express material shipments to meet the accelerated schedule, additional safety measures, and so forth. However, caution should be exercised in the calculations to avoid double-dipping. For example, express shipments are more costly, of course, but the lesser cost of normal shipments must be deducted.

Escalated labor costs: Labor costs frequently rise during the course of construction, and, when a project is delayed, the delayed labor for work in earlier periods is performed later at a higher labor rate. The differences must be demonstrated and included in the claim. For this purpose, it is useful to have a construction schedule that was submitted with the tender or soon after signing the contract, and which includes manpower loading. It is important to remember that labor escalation costs may be incurred even though the construction schedule is accelerated to meet the original completion date.

Escalated material costs due to later shipments: When a project is delayed, material shipments can either be delayed, at the risk of escalating prices, or they can be received as per original schedule and stored for the delay period. Material costs may also rise if extended warranties are involved.

Additional allocated general overhead: Even though general overhead has little to do with a project (whether delayed or not) it must be allocated to all the projects to be reimbursed. There are various ways to calculate this additional cost (see Chapter 3 for details).

Lost profit: Lost profit occurs as a result of the inability to obtain and service new revenue. Chapter 3 gives examples and illustrations of the basis for this charge. It should not be considered a form of gouging the respondent but, rather, a type of cost to the claimant.

Additional taxes: Any or all of the additional costs may attract additional taxes, either local or federal or both, and the claim should identify these taxes in this section.

The quantification summary: All of the above cost categories are listed in summary form. This listing can be done either at the end of this claim-submission section or in its own section under a separate tab. Details of calculations need not be listed here; it is sufficient to refer to the section where the detailed calculations can be found.

THE TOTAL AND MODIFIED TOTAL COST METHODS OF ESTABLISHING THE DAMAGES

Some contractors, and some lawyers and claims experts, feel that trying to establish detailed calculations for their delay damages is a waste of time. Instead, they compare the estimate with the total marked-up cost of the project, after completion, and submit their claim for the difference. Alternatively, they may make adjustments for any known shortcomings, on their behalf, and submit a claim for the difference of this modified total cost and the estimate. If this method is chosen, it is usually wise to be prepared to prove to the recipients of the claim that the estimate is reasonable and fairly accurate.

THE APPENDIX OF REFERENCES

The Appendix of References is the last section of the claim submission. It can be as short or as long as one wants to make it, but, except for invoices and time cards, the recipient usually has the same information as the claimant, and the preference is to keep it short. The main portion of this appendix consists of references to some written documentation (mostly correspondence, often several pages long) and, rather than include copies of these documents, I prefer a tabular summary (see Table 5-2) that gives the date of the document, the parties who initiated and received it, a relevant excerpt from the document to which I want to draw attention, and a column for remarks. Even this short form of documentation reference is often several pages in length.

XYZ PROJECT – Documentation Reference

DATE	FROM	TO	EXCERPT	REMARKS

Table 5-2 Excerpt summary of various documents

A similar table can be provided for references to independent authorities. It is often advantageous to provide such authorities to back up one's contentions, conjecture, and damage calculations throughout the claim submission, and the detailed references are then provided in this appendix in lieu of endnotes.

The method of supplying summarized tables not only cuts down on the bulkiness of the claim submission, but it also provides a quick overview of the most important documents. Occasionally, it may be necessary to include a copy of an entire document, especially when important issues must be brought out, such as marginal notes, signatures, and so forth, which are difficult to convey in an abridged format.

Invoices and time cards are usually not included in the initial claim submission; however, they may be required to be submitted at a later stage. Therefore, it is wise to treat them similarly to all other evidence, especially since the claim may proceed to arbitration or to trial. It is always wise to assume that the recipient may wish to verify the claimant's assertions, and to prepare for this eventuality. Furthermore, the claimant should be prepared to submit not only the invoices related to the delay but also those related to the entire project if productivity losses are at stake, since the defendant may wish to compare the invoices with the estimate to assure himself that the losses are not due to estimate shortages.

WHO SHOULD PREPARE THE CLAIM SUBMISSION?

If the claimant has an in-house claims expert, this question is an easy one to answer: it should be the in-house claims expert. However, this person, too, needs help from the project manager and/or field supervisor(s) who were involved with the project. On the other hand, if the claimant has no in-house claims expert, the project manager and/or the field supervisor(s)* must prepare the claim submission, preferably with the guidance of an independent claims expert.

Some claimants prefer to have an independent claims expert prepare the claim submission. This practice has advantages and disadvantages. First, an independent claims expert requires the assistance of the project manager and/or field supervisor(s) in any case, and, second, if the independent claims expert is too involved with the claim submission he may not be able to act as an expert witness, because the defendants and the courts will consider his testimony tainted or biased. Therefore, if an independent claims expert prepares a claim submission, it must be in the form of an opinion rather than as a statement of facts, which can only come from the testimony of those who were directly involved with the project, that is, the project manager and/or the field supervisor(s).

It is best, therefore, to have the project manager and/or field supervisor(s) establish the facts in the form of a claim submission and to have the independent claims expert give his opinion based on these facts. This procedure avoids the suggestion that the independent claims expert's opinion was tainted by the assembly of facts that are not supported by his direct knowledge of or involvement with the project. The independent claims expert is usually in the best position to render an unbiased opinion

---

* I mention both the project manager and the field supervisor(s) because there is an essential difference in their endeavor to get a project completed: field supervisors have a tendency to ignore or sacrifice their employer's legal rights and obligations under the contract in order to get the job done, whereas project managers are constantly aware of and fighting for them. It is a question of focus.

based on the facts if he had little involvement with their assembly. Should it still be necessary to have an independent claims expert prepare the claim submission, it would be advisable to have a second independent claims expert render an opinion.

GETTING THE OPINION OF A CLAIMS EXPERT

Regardless of whether or not a claims expert's assistance is required for the preparation of the claim submission, a claim expert's opinion on the value of a claim should be sought. A claims expert who is familiar with the pros and cons of delay claims can usually render an opinion on a particular set of circumstances in a relatively short time period and with minimum expense. Sometimes the facts do not support a good claim, or the facts may be too skimpy to support a good claim, and the claims expert will point this out to the claimant before needless expense is incurred. Also, a poorly prepared claim reviewed by a claims expert can still be corrected, which may be difficult to do after it is submitted.

GETTING THE OPINION OF A CLAIMS LAWYER

Similarly, it is wise to obtain the opinion of a good claims lawyer who is familiar with delay claims before proceeding with the claim submission. A lawyer will look at the facts from a legal viewpoint when rendering his opinion. For example, an experienced lawyer will not only consider the ramifications of possible contract breaches but also whether or not there is a case in tort. That is not to say that the lawyer is always right — you have, after all, lawyers on both sides who disagree — however, if your lawyer is honest with you, you will at least find out the legal downside to your claim before you submit it.

RECEIVING AND RESPONDING TO A CLAIM

The usual, initial reaction, when one receives a claim, is to reject it outright, with feelings of disgust and dismay. Paying for a claim, even when it is justified, is like purchasing something of no value.

Delays and disruptions, regardless of who caused them, usually affect everyone financially, and the party at fault has probably shown tolerance in the past with other parties who caused them. Nevertheless, an outright rejection of the claim is not a solution to the problem — at least the damaged party will not accept it as a solution but will probably proceed to a legal remedy.

A better response is to be noncommittal but show interest. First, go for information. Information must be produced during the discoveries before a trial, and claimants will seldom refuse to produce it sooner upon request from the recipient, because supplying the information may help a settlement without having to resort to legal remedies. I asked a friend once what is the first step he takes when he receives a claim from a subcontractor — he was, at that time, a project manager for a national contracting company. He said he usually contacts the claimant immediately and requests his estimate, his time sheets, and his invoices for the project, telling the claimant he cannot properly assess the validity of the claim without them. This request is a good first step, in my opinion. A claim is seldom submitted without a good cause, but the claimed effect is another matter.

All too often, a contractor will blame his entire loss on a project on the cause that resulted in the claim. Even when he modifies his claim quantification by some of his own shortcomings, he usually leaves the recipient of the claim with a feeling of being ripped off, shortchanged, as it were, which may or may not be the case. Therefore, getting more information before responding to a claim is always a better idea than jumping to wrong conclusions.

CHAPTER 5 REVIEW

A claim submission usually starts with an executive summary that gives a brief overview of the causes and effects of the claim. This overview is followed by detailed descriptions of the causes and a more detailed overview of the effects, as well as any mitigating measures that were taken to reduce the impact of the effects. The next section of the claim submission gives the details of various damage quantifications or, alternatively, reasons why the claim is based on total, or modified total, cost overruns. Any references required for, or mentioned in, the various claim sections are usually included in an appendix.

It is advisable to prepare the claim submission in-house — preferably by the project manager and/or the field supervisor(s). Nevertheless, opinions from a claims expert and a claims lawyer should be obtained before submitting the claim.

The recipient of the claim usually feels like rejecting it outright, but the better course of action for a recipient is to go for more information and to keep emotions out of the response.

6
Settlement Negotiations
and
the Alternatives

GENERAL COMMENTS

A claim can be settled in a number of ways (mediation, arbitration, or the judicial process) but, all things considered, negotiation is probably the most sensible way — providing, of course, that both parties are willing to negotiate in good faith. In the extreme alternative, when people refuse to negotiate, or even talk to each other, and the claim proceeds to trial and to appeal courts, the process takes many years at astronomical personal and financial costs.

Negotiations are often believed to be synonymous with compromise. Each party believes that the evidence favors it, so why should it negotiate and compromise its position? Sadly, this belief is often supported by the claims expert and/or legal counsel, seldom selflessly; it also creates a tendency toward inflexibility. Therefore, a good initial step is to convince the other party that negotiations will benefit both, because judges or arbitrators will seldom accept or reject all of the claims of one party, which in itself leads to compromise.

Keep in mind that the party who is asked to pay for a claim buys nothing tangible; you are up against the problem of convincing the recipient, who may think that his part in the claim causes were minor, to pay for all the damages without a benefit in return — no *quid pro quo*, as it were. The trick is to find solutions in which everybody wins. This endeavor is by no means easy, but if both parties sincerely put their minds to work to achieve this outcome, it may just be possible. As an example, if the claim is by a contractor against an owner who is going to order more construction of future projects, the contractor could suggest absorbing some or all of the damages in return for a future contract — at a reasonable price, of course.

NEGOTIATIONS AND SETTLEMENT DURING CONSTRUCTION

The best time to settle a dispute is during construction: the evidence is obvious and the staff involved is still available. If claimants make it their policy to negotiate disputes as they occur, they will automatically give proper notice of these events. Also, the costs of the damages are usually less disputable, and mitigation by both parties is still possible.

A frequent problem is that many delays are of a minor nature as they occur, and the parties affected are reluctant to cause a fuss over them and are willing to absorb the additional costs. It is not until it becomes evident that the cumulative effect of these so-called minor delays and disruptions has a much larger impact that notices of claims are issued. This realization can happen near the end of construction when it is usually too late to correct the situation or to mitigate the damages; consequently, the respondent to the claim is extremely reluctant to admit responsibility and to negotiate a decent settlement.

Another advantage of negotiating and settling delay claims as they arise during construction is that the people most intimately involved with the causes and effects, the project manager and the field supervisor(s), can be called upon to voice their concerns and to provide fresh evidence. However, it is wise to leave settlement negotiations to top management, because field supervisors have a tendency to be more lenient in an effort to get the job done at all costs. Also, field supervisors may not have a good grasp of the total impact of the delays and disruptions involved.

Since contractors and owners are usually reluctant to offend each other by causing a fuss over so-called minor incidents, many chances are lost to correct problems that eventually cause major impacts. In my opinion, claims could actually be avoided if the parties *would* cause a fuss over so-called minor delays and disruptions, because they could then take the opportunity to make corrections and to avoid recurrences — even major delays and disruptions can often be corrected or properly mitigated if caught in time.

Notwithstanding that negotiations during construction may turn out to be fruitless, I would still conduct them on a "without prejudice" basis, since this very process will force everyone to keep records of the issues involved, which can later be produced if a claim is litigated, and some issues can either be immediately clarified or even resolved.

NEGOTIATIONS AND SETTLEMENT AFTER CONSTRUCTION

Negotiations after construction usually occur if a claim submission was delayed till the end of construction or if one or both of the parties refused to negotiate a settlement during construction.

There are several reasons for a claim not being submitted until the end of construction: the most likely are that either the claimant could not establish his delay and disruption costs any sooner or he did not decide any sooner to submit a claim, perhaps because of gradually accumulating causes that eventually became too much.

There could also be several reasons that one or both of the parties refused to negotiate a settlement during construction. An owner, for example, may wish to find out first if he is satisfied with the contractor's performance before committing himself to an extra outlay that, to his mind, may amount to nothing more than a bonus. A contractor, on the other hand, may be afraid that claim causes are still accumulating and that signing the inevitable release for an early settlement may jeopardize further claims.

Whatever the reasons for delaying negotiations and settlement until after construction, this timing is usually worse for a contractor than for an owner. The old adage that possession is nine points of the law applies here. Once an owner is in possession of a finished facility, he can just refuse to pay any more for it than the contract amount. In fact, some owners even refuse to pay the contract holdback amount until all liens *and* claims are removed. This action alone can take substantial litigation to overturn. In any case, before finishing construction, a contractor still has some clout in his negotiations for claim settlement. Nevertheless, the clout is weak, because his contract is usually specific about his performance, and he could find himself in breach of contract if he is not careful.

Another disadvantage of negotiating a claim settlement after construction is that most of the physical evidence, and sometimes the people who were involved, have disappeared, and the claim relies mainly on documentation, good or bad, both for the claim causes and for their effects. However, good documentation is often missing for the same reasons that negotiations were delayed in the first place.

NEGOTIATIONS AND SETTLEMENT THROUGH MEDIATION

A mediator, as opposed to an arbitrator, can only make recommendations to the parties, not binding decisions. So, why get him involved? Well, the parties may be driven to settle the claim through negotiations rather than through litigation, but emotions and tempers may prevent them from doing so: they may not even wish to talk to each other any more. A mediator, in such circumstances, can serve as a go-between.

But the role of a mediator can be much more useful. He can listen to both parties' logic and arguments and patiently point out the flaws in them. The parties are much more likely to listen to a neutral person, the mediator, about flaws in their cases than to the opposing side, whom they rightly suspect of being prejudiced. Furthermore, prejudiced people often have a mental block that is not even recognized by them, and one prejudiced side is not likely to take advice from the other prejudiced side about removing such a mental block. However, these parties may accept such advice from a mediator as long as they believe the mediator is fair.

A mediator can also bring an inventive side to the table. He can often see solutions that can assist in breaking deadlocks. The solutions may even benefit both parties. If either party had proposed such a solution to the other party (without the assistance of a mediator), the other party may have viewed the proposal with great suspicion. The problem is that neither party may have proposed, or even thought of, such a solution because of its mental blocks.

For these reasons, a mediator can be a useful catalyst during negotiations. It may be a good idea to agree *before* starting negotiations to involve a mediator, when the parties are still agreeable with each other. However, a party may object to getting a mediator involved for fear of compromising its position. I have run into this situation and, knowing how valuable a mediator can be to help resolve deadlocks, I have suggested the use of a person from the opposite camp who was not involved in the direct negotiations nor in the previous, heated discussions, and whom I trusted to be reasonable. This tactic does not always work, because such a person still brings bias to the table, but it should be kept in mind as a viable alternative to a neutral mediator.

In some cases, a mediator may make written recommendations for the parties to consider. This practice is useful when people have stopped listening. The proposals can then be studied at leisure and submitted to higher authorities who were not part of the direct negotiations but who got their information secondhand. As mentioned earlier, none of the recommendations of a mediator have to be adopted by the parties, and this aspect, too, is one of the advantages of mediation.

BEHAVIOR DURING NEGOTIATIONS

Claim negotiations are a little different from ordinary negotiations. In ordinary negotiations, one party usually has something to offer that another party wants — such as selling an item or a service — but in claim negotiations, one party wants to be reimbursed for costs that it should not have incurred and the other party not only refutes the claim but also feels it is paying for something of no tangible value. This dilemma makes negotiations difficult at the best of times.

The claimant can, of course, take the position that the claim is solid and well founded, and that, if the recipient refuses to recognize this fact and refuses to settle the claim voluntarily, he can always litigate. This attitude works sometimes, but more often it antagonizes the situation further. The problem in construction is that nobody's slate is ever clean, and claim recipients know this and make the most of it. I have watched claim recipients shoot all kinds of holes in "solid" claims because the claimant had a few minor performance problems. These reasons are why claim negotiations should be based on different strategies from ordinary negotiations and why you should follow three main principles:

1. Do a sincere and honest job of explaining to the claim recipient what caused the extra costs, how the extra costs were arrived at, and what was done to keep the extra costs to a minimum.
2. You should not get too aggressive about your position, even when you feel you can win your case in court. A meek, almost hat-in-hand, attitude will serve you better; after all, if this tactic does not work, you can always proceed to litigation without having threatened to do so. The claim recipient probably knows this, too.
3. Keep stressing the business relations angle. Remember, it is a long road that does not have a turn in it somewhere, which is especially true in construction: contractors are always looking for more work, and owners usually want to build more facilities or expand existing ones. Besides, word of a hard-nosed position gets around pretty fast, which may hurt you in the long run.

When negotiations are done openly and with due consideration for one's own and the other party's weaknesses, as well as with the future in mind, a successful outcome is much more assured. Try to be civil about everything and keep personalities out of it; if this approach fails, there is still time to consider other remedies.

SETTLEMENT THROUGH ARBITRATION

If negotiations fail, arbitration should be considered. Arbitration can be entered into with or without the help of legal counsel, but legal counsel should be consulted regarding the advisability to proceed without counsel. Some contracts do not allow arbitration and other contracts allow nothing but, and, when arbitration is an option, the parties should still be advised if arbitration is the best course of action under the circumstances. Keep in mind that arbitration is usually binding on the parties, and often the rights of appeal are limited and the process is virtually always expensive.

I mentioned that the parties could proceed without the help of legal counsel, but, unless the case is simple and open-and-shut, it is not advisable. Arbitration usually requires each party to present its case, including witnesses, if any, and then the arbitrator considers the facts presented and renders his decision. If anything was missed during the presentation, too bad. Arbitrators do not normally initiate further investigations into, or searches for, additional facts. This limitation makes the help of legal counsel valuable, because he is trained to present all relevant facts to people like judges and arbitrators.

Arbitrations can be handled by a single arbitrator, but the usual process is for each party to appoint one arbitrator and for the two party-appointed arbitrators to choose a third arbitrator who usually acts as the chair. All three arbitrators are expected to be neutral and fair, but the party-appointed arbitrators usually come to the board with at least the perspective of the party appointing them. This bias can lead to a tie between the party-appointed arbitrators, which is broken by the chairman. The party-appointed arbitrators should refrain from getting any information before the arbitration board receives the facts, to leave themselves in the best neutral position. Otherwise, one or the other party-appointed arbitrator could find himself in possession of information that was not presented to the entire board; this knowledge would not only create a bias in him but may even show him to be biased and could be grounds for misconduct, leading to dismissal.

The question then arises, why not have just one neutral arbitrator? The answer is simple: the claimant and recipient may have substantially different backgrounds and will, therefore, stress different concerns in their presentations. If each of them appoints his own arbitrator, he will choose a person who is most familiar with his particular concerns and circumstances and who can convey these concerns to the co-arbitrators. If only one arbitrator is chosen for both parties, he may have leanings to either one or the other's concerns and circumstances.

Despite the unlikelihood of appeal ability, there is much to be said in favor of arbitration. The chosen arbitrators are normally familiar with the construction industry; they not only bring along empathy from their own experience but also readiness to understand the issues. However, the arbitrators may not be too well versed in the legal issues involved, and this lack is just another reason for being represented by legal counsel.

Negotiations during arbitration proceedings are almost nonexistent, partially because of the missing discovery process required for trials, and partially because the arbitration time frame is much shorter than for trials. Whatever the disadvantages of the arbitration process, the shorter time frame and the experience of the arbitrators should be weighed in the balance. The process of educating lawyers and judges to understand the specifics of the case and the construction industry can be time consuming and costly. Also, the courts are usually overloaded, and complex cases can take many years to resolve, especially if they are taken through the appeal processes. Arbitration, on the other hand, seldom takes more than a few weeks or months to conclude and costs less than the cheapest trial.

OTHER ALTERNATIVES

A number of other alternatives are available, including judicial dispute resolution (JDR) such as pre-trial conferences, judicial mediations, expanded pre-trial settlement conferences, and mini-trials. Most alternatives are only minor variations of mediations or arbitrations, except for one variation of the mini-trial, which I shall describe in more detail in the next section. Nevertheless, each of the alternatives may offer a unique solution for a particular circumstance and should, therefore, be carefully considered.

THE MINI-TRIAL

Mini-trials vary across the country and depend somewhat on the procedures established by the member of the court who will conduct the mini-trial and the agreement of the parties. I will endeavor to describe one version of the mini-trial that I consider useful to the process of negotiating a settlement without resorting to a full-blown trial by court.

Mini-trials are normally conducted by a judge, either an active one, or a retired or semi-retired one, who can sometimes be supplied by private, alternative dispute resolution firms. Counsel are usually asked to prepare presentations similar to their closing arguments at the end of a formal trial. As much as possible, counsel normally agree on the facts. At the mini-trial, with their clients present, counsel will then be asked to tell the judge in summary form what they would expect their witnesses to say at a formal trial, and to produce summaries of any written materials that would be tendered as exhibits, including experts' reports. The judge may ask counsel questions to clarify issues and, at the conclusion of counsel's presentations, the judge may ask their clients if they have anything to add. Giving the clients an opportunity to speak assures that they get anything that is bothering them off their chests before a judge; this opportunity makes them part of the trial process. The judge may also ask counsel and their clients if they agree that he knows as much about the case as the trial judge would after a formal trial hearing. The mini-trial judge will make sure that the parties know, should they decide to proceed to a formal trial, he will never be the judge.

The proceeding is usually amicable. At the end of the mini-trial, the judge retires to consider the issues and then returns to give his opinion. The judge explains how he would decide the issues and tries to give his reasons for coming to his conclusions. Since mini-trials are usually non-binding on the parties, although they can be binding by agreement, the judge, without putting pressure on the parties to settle, usually points out that they can accept or reject his conclusions, and, if they reject them, they can proceed to a formal trial, usually at enormous cost, stress, and delay, and without guarantee that they will end up satisfied. Counsel will sometimes convince their respective clients to accept the judge's conclusions as binding, but, more likely, the parties review their positions in light of the judge's conclusions and reach an agreeable settlement.

The advantage of this kind of mini-trial is that it gives the parties a feel for what may happen to their case at a formal trial. It is like getting a neutral expert's opinion. When the weaknesses of their respective cases are pointed out to them by a judge similar to the one they can expect at a formal trial, they are more likely to pay attention and take the opportunity to reassess their positions, without having to spend time, energy, stress, and money on a formal trial that may conclude similarly. The mini-trial is especially useful to determine a point of law, a contract interpretation, or other issues that are not hotly contested.

The mini-trial process is informal. Materials and submissions are strictly confidential and for the purpose of the mini-trial alone. Reference may be made to applicable authorities, to experts' reports, and to evidence taken at examinations for discovery, if any, but, at the mini-trial itself, evidence is normally not produced. Nevertheless, the process is not suited for every lawsuit. Cases where facts are in dispute or where the issues turn on the credibility of conflicting evidence are usually not suited for mini-trials. Furthermore, some construction disputes may be too complex to properly submit in abbreviated form at a mini-trial. The exception, in the latter case, occurs when the parties can agree on a summary of essential facts to be submitted to a judge, even though they may still have some conflicting views on these facts.

Sometimes, a formal trial can inadvertently turn into a mini-trial. I know of one case where the judge, halfway through the trial, called counsel to his chambers and asked them if they would like to hear his opinion now or wait till the end of the trial. The lawyers opted for an immediate opinion and then went to their respective clients and convinced them to settle out of court. It is not unknown for judges to be interventionists if the issues warrant this action.

DISCOVERIES

Discoveries are a relatively modern procedure to cut down on court time and expense. They take place before trial, and are sometimes referred to as examination before trial (EBT). Other terms in connection with discoveries are depositions and interrogatories. The rules of court differ slightly in various jurisdictions, but the procedures are similar. There are usually discoveries of documents and discoveries of witnesses. When witnesses are involved, the usual procedure is the oral examination for discovery, but there can also be written examinations, which are referred to as interrogatories.

For the discovery of documents, which is usually completed before oral or written examinations, the litigants produce for their respective lawyers all copies and originals of documents (including computer records) relating to the lawsuit as well as a list of documents they no longer have and the reasons for the absence. The parties may consider some of the documents privileged and should tell their lawyers why they should remain so. The lawyers will then decide which documents must be made available to the other parties in the lawsuit. A list or affidavit of all the documents is then compiled, and each litigant must swear under oath that he has produced or described all documents required. These will be the only documents the parties can use at trial.

The next stage is the oral examination of witnesses. This process can take place in either of the lawyers' offices or in a court reporter's office. The procedure is informal: usually only the litigants, their lawyers, and the court reporter are involved. This informality can be deceiving, because the witnesses must swear under oath that their answers will be the truth, the whole truth, and nothing but the truth. They should keep in mind that the discovery is an official stage in the lawsuit, and anything they say can be used against them. Depending on the issues, the examination can be as short as an hour, or it can take several days. Witnesses must make themselves available (attendance is mandatory) until the examiner has completed his examination, usually after the witnesses are re-examined on any undertakings they had to supply.

Sometimes, one of the parties may issue interrogatories — a list of written questions served on the adversary, who must supply written replies under oath. Interrogatories are not as flexible as oral depositions, which can be cross-examined, but they are almost as good and fairly inexpensive to establish important facts held by the adversary. The main advantage of oral examinations is that the behavior, credibility, and value of witnesses can be assessed by either side before the case proceeds to trial; furthermore, additional facts that were not conceived for interrogatories may surface.

Some of the trial witnesses may be expert witnesses, who may or may not be accepted as such by the opposite party. If an expert witness raises questions in the minds of the opposition, or if the expert witness is not available to attend the trial, a deposition may be taken; if this practice is allowed in your jurisdiction, it can be an effective tactic to get the lawsuit resolved.

In most cases, witnesses are briefed by their lawyers, and it may be redundant to say much more about witnesses' behavior. However, since a great deal depends on effective witnesses to resolve a lawsuit, it is worthwhile to discuss the behavior of witnesses who make a difference, regardless of the value of their testimony. Everyone who will ever be called upon to be a witness should keep a few basics in mind. An unprepared witness can ruin an otherwise good claim.

WITNESSES WHO MAKE A DIFFERENCE[*]

Construction claims usually involve companies, and companies usually ask employees, independent experts, and an officer to act as witnesses on their behalf. Nevertheless, any employee or former employee may be examined by opposing counsel. Only the examination of an officer is binding on the company, but keep in mind that this person may not make the best witness or may not be familiar with all of the facts. Some witnesses are themselves party to the litigation, and others are independent, that is, bystanders who became witnesses. Occasionally, a chosen witness may be reluctant to be examined, and a subpoena may be necessary requiring the attendance, subject to penalty for noncompliance. Companies will, of course, try to choose someone who is most familiar with the facts of the case, but this person may not make the most effective witness.

The first preparation is for a company witness to familiarize himself with all the facts, even if he has personal knowledge of the case. This is a typical situation in which a little bit of knowledge can be a dangerous thing. The next step is for the witness to be briefed by the company's lawyer to better understand a witness's function. This practice has the added advantage of informing the lawyer of the witness's knowledge. Witnesses who prepare themselves in this way have a better chance to do well. It is best to prepare shortly before discovery — a mock examination may be useful. The witness should organize his thoughts about the issues in the lawsuit, but, with the exception of written records, he can rely only on easily remembered mental notes rather than on written ones. The witness should also be prepared to answer personal questions about himself: his experience, employment history, education, military service, and so on, all to flush out any potential skeletons in the closet.

One thing to keep in mind is that an examination for discovery is not a conversation, even though it may appear that way; it is a question-and-answer session, with the answers given under oath. Be as brief as possible with your answers: in most cases a simple "yes" or "no" suffices. Some witnesses feel they must explain themselves; remember, you do not owe anyone an explanation, so do not volunteer one. Witnesses who become defensive during discovery make themselves look guilty. Remember, you are only there because of what you know, and the

---

[*] It is not within the scope of this book to provide detailed briefing. A good book is available for this purpose, written by Stuart B. Shapiro, titled *How to Survive a Deposition*, published by John Wiley & Sons, Inc., 1994.

more information you add to an answer, the more you run the risk of volunteering information that should not be volunteered. Providing additional information will only open the door for the questioner to go on further fishing expeditions. Remember detective Joe Friday's words: The facts, ma'am, just the facts.

On the other hand, do not withhold any information that would make your answer complete. Be truthful no matter what. Truth has a way of coming to the surface, and once you are caught being less than truthful, you lose all credibility as a witness, and the party you represent may even lose its lawsuit. Never be afraid of the truth; what may appear to be harmful could well turn out to be useful. And do not try to outfox the questioner by guessing at his objective; this practice will only lead to providing answers that are partially untruthful or evasive.

Also, do not be afraid to answer "I don't know" or "I can't remember". These are far better answers than guessing or being evasive. We are sometimes ashamed to admit we do not know something that we should know, or that we cannot remember it, but these are the facts of life: better to admit it than to offer a guess that could only hurt the lawsuit in the long run. Remember that it is a violation of your oath if you say "I don't know" when, in fact, you do know. In the same vein, avoid speculating. For example, do not offer "I'm not sure, but I think...such and such". Also, do not be afraid to say "I don't understand the question" when you do not understand it. Many times, lawyers ask complex questions; request that the question be rephrased. Complex situations usually become simple when viewing the details. Also, beware of compound or double questions; you may unintentionally answer the wrong part of the question. It pays to ask the examiner which part of the question he wants answered. Never answer a question you do not completely understand. Otherwise, be positive and do not waver. And keep in mind the *written* record you leave behind: remember, voice inflections and body language do not end up in the transcript.

A good witness will be natural and maintain his composure, even though the examiner appears to be hostile. Do not start arguing with him. Pause briefly before you answer; besides giving yourself time to reflect on your answer, you give your lawyer a chance to voice an objection. Should he do so, stop immediately and listen carefully; he may be providing you with a hint to be careful, or he may only object to test the skills of his opponent. Speak clearly and slowly, preferably facing the court reporter, who must sometimes revert to lip-reading. Do not discuss the case with anyone but your lawyer — especially before discovery. If you did comment about the case to anyone, let your lawyer know.

You will get a chance to read the transcript of the examination for discovery soon after the discovery. This is the time when you should let your lawyer know if corrections are necessary. Make sure before the discovery that your rights to make corrections have not been waived. You must keep in mind that, in most court jurisdictions, the opposing lawyer is the only one entitled to use the transcript at the trial; you should, therefore, make sure your answers are correct.

A well-prepared, composed, and credible witness can often make the difference between settlement during or immediately after discoveries and letting the lawsuit proceed to trial. Every effort should, therefore, be made to produce and educate credible witnesses.

PROCEEDING TO TRIAL

When all of the pleadings and affidavits setting out producible documents have been filed, undertakings provided, and discoveries concluded, the parties may apply to have the matter set down for trial. Considerable disadvantages and risks must be weighed, however, before deciding on this step.

One thing to remember is that you can pick your lawyer but you cannot pick your judge. Judges are human beings, too, with their own philosophies, viewpoints, and experiences. If you get a judge who does not understand the issues, or a judge who has some unknown leaning, you may be better off settling out of court. However, if no settlement was reached during or after discoveries, when the weaknesses and strengths of the case should have become evident, a settlement immediately before or during the trial may also seem unlikely. Nevertheless, a settlement should still be attempted.

It is important to realize that both parties are at considerable risk when proceeding to trial. The most obvious risk is losing, of course, but there are other risks as well. Although the successful party may recover some legal fees and all reasonable disbursements, which can be considerable for an expert's evidence, this recovery is seldom enough to pay for all of the expenses incurred. Furthermore, when a formal offer of settlement has been made and the winning party does not achieve more than the offer, it will be penalized in costs.

Then there is the risk that a trial scheduled for a set number of days will take two or three times longer, at proportionately increased costs. Furthermore, trials are very stressful and require much time and energy from the parties, whose attention is taken away from more productive things. It takes years sometimes to advance a claim from commencement to trial, and there is the risk that in the end there may be insufficient funds to satisfy the judgment; this risk should be assessed at the outset.

The type of evidence available should also be considered. If the parties have insufficient documented evidence to support their claims, the judge must rely on witnesses who give oral and, perhaps, conflicting evidence, in which case his leaning will be toward the most credible witness, regardless of what the actual facts might have been.

Another real risk for the plaintiff is that he, himself, may have been responsible for contract infractions — perhaps these were missing or untimely notices of claim causes — which may attract a counterclaim. Do not shrug this possibility off with an attitude that it is only a scare tactic that the judge will surely dismiss. Remember, nobody's slate is ever clean. One never knows all of one's sins until one reads about them in a counterclaim. At the very least, a judge may wonder about the credibility of the parties and to what extent each party caused its own demise. I recall one such case where both claim and counterclaim were dismissed.

Finally, there is the risk that the plaintiffs' quantifications of damages may not be accepted by the judge, even if the liability issue is decided in their favor. Judges scrutinize and assess damage claims carefully, especially for validity, accuracy, reasonableness, and duplications. A badly inflated claim can easily ruin the credibility of the claimant and can even result in a judgment lower than it should be.

Judges know about these various risks and often explore the possibility of settlement at a pretrial conference; they may even recommend to the parties to use judicial dispute resolution for the purpose of settling the claim without a trial; and if settlement is not a possibility, the case is referred to a civil trial coordinator, who assigns a trial date.

At the trial, all of the documented and oral evidence is presented. Witnesses usually give oral evidence, but experts can submit written reports if the facts in their reports are relatively straightforward and uncontradicted. Pretrial applications can be settled on the basis of affidavit evidence — again, when the facts are not much in dispute. One example that comes to mind occurred when a contract holdback was not released because of the filed claim. The judge, in this case, ordered the release of the holdback, because retaining it was only a pressure tactic by the defendant and had nothing at all to do with the claim.

After the evidence is in, each party's lawyer is given the opportunity to present arguments and references to similar decided cases. The judge then takes the matter under advisement and, after carefully considering the evidence and arguments, renders his judgment. However, this judgment may not yet be the end: the loser's counsel will likely assess whether legal principles have been misapplied or evidence misconstrued; should this be so, the case will likely be appealed.

CHAPTER 6 REVIEW

The best time to negotiate and settle a delay claim is when the delay occurs, during construction. However, claimants seem to be reluctant to approach the problem during construction, either because the individual claim causes are minor or because the exact damages are too vague. Nevertheless, if negotiations proceed promptly, there is a better chance that proper notices are given and that good records will be kept.

Leaving negotiations until construction is complete can be detrimental, especially to a contractor, even though the costs are final. Staff may disappear, and the evidence is not as fresh. Also, the staff may have been too lax in issuing proper notices, assembling good documentation, or mitigating the effects.

If the parties are too upset to make effective dialogue possible between them, they may consider the use of a neutral mediator. But keep in mind that the mediator can make only recommendations, not binding decisions.

In many respects, claim negotiations are similar to ordinary negotiations, except that one party requests payment from the other party for costs that should not have been incurred and are of intangible value. This dilemma usually makes negotiations harder, and the claimant must be sincere and honest when explaining the causes and effects of the delay. A hat-in-hand attitude is normally more productive in these situations, and future business relations must also be kept in mind.

If negotiations fail, arbitration or some other inexpensive alternative should be considered. Arbitrations can be conducted with or without counsel, but there are usually enough legal implications to make counsel advisable. Furthermore, arbitrations are normally binding unless the parties have stipulated otherwise.

Other alternatives are usually only variations to mediation and arbitration, except for the mini-trial. In the mini-trial, a judge considers the evidence and renders an opinion, usually nonbinding. This process gives the parties a good idea of the possible outcome of a formal trial.

Should none of the alternatives be acceptable, the matter usually proceeds to discoveries, in which both parties have an opportunity to find out each other's strengths and weaknesses. Good witnesses can make a tremendous difference and can lead to early settlements.

If the matter proceeds to trial, the parties must carefully consider the downsides and risks involved. The process is lengthy, costly, and stressful, and even the "winner" will not recover all of his costs. And, if the trial decision is appealed, more costs and stresses are involved.

7

Summary, the Business Relations Angle, and Conclusions

GENERAL COMMENTS

In Chapters 1 to 6, I wanted all parties involved with construction and construction disputes to take a *new* look at construction estimates, at measuring job progress and achieved productivity, at the difference between task postponement and project delay, at effective reporting procedures, at calculating and allocating overheads, at potential duplications between interest charges and return on investment, and at the risks of litigation. A new look may result in better practices, with a consequent saving of time, money, and anguish.

Some owners, contractors, and designers have the conviction that learning about delay claims may cause the number of claims to increase, but I believe that a good understanding of the underlying causes and all of the concepts involved in establishing delay damages will actually lead to a decrease of delay claims. Moreover, a good understanding of the causes and costs of delays and the risks of litigating a claim for the damages must surely lead to the desire to try to eliminate such claims; when this elimination is not always possible, the various groups involved in construction can still be helpful in mitigating the effects of delays and, perhaps, in seeing to it that the party most hurt is properly compensated.

Following is a summary intended to give an overview of the various subjects treated in this book.

SUMMARY

Construction delays can often be anticipated. Chapter 1 related delays to the original estimate and advocated that the estimator should include delay costs, rather than hope for the best and be faced with a later claim. The establishment of a preliminary construction schedule to be included with the tender is also recommended. This schedule not only helps the contractor with his estimate, but it also confirms his intent about starting and completing various phases of construction. Since change orders to the contract can be disruptive to the original contract work, quotes for changes should be specific regarding the expected impacts and delays to the project. The financial benefit of change orders is often overestimated by contractors, owners, and designers, all at the expense of underestimated delay and disruption impacts on the total construction work and the possibility of failing to complete on time.

Chapter 2 showed us, graphically, delays that occur at various phases of construction and also showed when delays affect the completion of the project. Then it explored various methods of assessing and measuring the achieved productivity and the effect of the learning curve on production. The task-per-area assessment of establishing fairly accurate percentages of progressive project completion was explored and contrasted to similar percentages of labor-budget expenditures to find the level of productivity. Next, the value of issuing weekly field progress reports was covered. These reports summarize the week's accomplishments and constraints as well as the work planned for the following week.

Chapter 3 described the various overheads encountered in the construction industry and how these overheads are calculated for, and allocated to, the individual projects. We were shown, graphically, why overheads are higher for delayed projects and to what extent these projects should receive higher general overhead allocations. We delved into the components of profit: the known and unknown risk allowances, and the return on investment. We were also given illustrations of how interest charges are calculated for delayed projects and when interest charges should not duplicate return on investment.

Chapter 4 delved into why the preparation of documentation is more valuable when it is completed *during* rather than *after* construction. We were also shown that assessing and allocating production losses is more accurate and effective when it is done at the time of occurrence; if one waits until construction is completed, one may have to resort to using the modified total cost method, which is often frowned upon. Good accounting methods were advocated to establish accurate delay costs. Furthermore, claimants and recipients of claims should be well versed in the contract conditions that cover schedule delays and entitlements for damages.

Chapter 5 described a method of preparing a claim submission, with detailed outlines of its various sections. It also discussed the advantages of preparing a claim submission in-house rather than by a claims expert. Nevertheless, this chapter recommended getting the opinions of a claims expert and a claims lawyer before submitting the claim.

Chapter 6 told us that the best time to negotiate a claim settlement is during construction, when the facts are still evident and are still fresh in everyone's mind. Besides, at this time contractors and owners are still in a balance of power. However, contractors are often reluctant, for various reasons, to engage in negotiations during construction. Whether negotiations are begun during or after construction, it may be advisable to engage a mediator, especially if the parties are too upset to talk to each other rationally. If negotiations fail, the parties may want to consider arbitration or some other inexpensive alternative to a trial to resolve the dispute — perhaps a mini-trial, which is not binding but tells the parties what they might expect at a formal trial. This chapter also explored discoveries, witnesses who make a difference, and the various risks and downsides of proceeding to trial.

THE BUSINESS RELATIONS ANGLE

As if the subject were taboo, the various parties in the construction industry are usually reluctant to talk about potential delay claims, or to give each other hints that they are even thinking about them, all for the sake of maintaining good business relations. The construction industry is, after all, relatively small, especially as far as the individual niches are concerned, and people realize that they will most likely work together again in the near future.

I think it is unfortunate that the subjects of delays, slowdowns, and disruptions, or even the possibility of them, and the potential damages they can cause, are not openly discussed, with the prospects of preventing them, or at least finding early remedies to mitigate their effects. If this discussion does not take place at the outset of construction, business relations are generally damaged much more severely at a later stage, when one of the parties finally decides it can no longer tolerate the situation or absorb the losses that have been incurred. By this time, it is usually too late to give timely notices and to do anything to eliminate the problem or to mitigate the damages.

However, keeping the potential benefits in mind, business relations should play a big role, in my opinion, when it becomes necessary to submit a delay claim. As mentioned previously, this is a good time for people to remember that it is a long road that has not got a turn in it somewhere. If everyone involved with the claim keeps this philosophy in mind and consciously applies good business relations principles, finding an acceptable solution will be much more likely, even at this late stage, and damaging future relations will be less likely.

CONCLUSION

Too little is known in the construction industry of the devastating effects of delays, slowdowns, and disruptions. You may ask, rightfully, if these effects are so devastating, why is not more known about them? The answer, I think, is because the effects are insidious: they occur mostly in smaller, neglected quantities, and their cumulative effects are not known until much later, usually too late to do anything about them except for trying to collect on a claim.

On the other hand, construction people who know what is in store if delays, slowdowns, and disruptions are allowed to continue, and try to forewarn others, are seldom appreciated, sometimes not even by the companies for whom they work. This lack of understanding, in my opinion, calls for widespread education in the construction industry. We need to be more aware of the various causes of the delays, slowdowns, and disruptions, and of the various facets that make up their devastating effects (the damages). Also, we need to know what delay claims are all about and how to process them.

This book is intended to assist in this task; it is not intended to advocate for, and increase the number of, claims. A good education of what delay claims are all about not only increases a person's awareness level but also his appreciation and his desire to avoid the pitfalls as much as possible − not to mention assisting others to do so.

In the final analysis, all of us in the construction industry should be interested in helping each other whenever possible to level out the playing field through knowledge, education, and communication rather than through isolated trial and error. The business relations factor is one good reason, but the main reason is that this industry can only become successful if all of us, rather than a few of us, become successful.

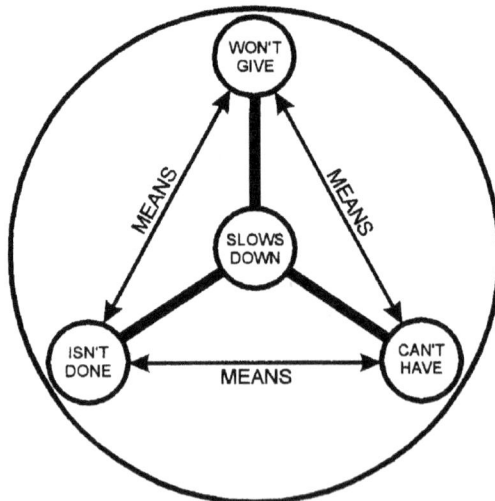

Appendix
List of Delay Causes
and
Production-loss Examples

GENERAL CONSIDERATIONS

Before we can talk about production loss, a fundamental understanding of productivity is essential — especially when trying to sell the concept in a construction-delay claim.

Is there such a thing as "normal" productivity? Let's look at two different machines in a factory, producing similar widgets: one machine is older and produces 90 widgets per hour; the other machine is more modern and produces 100 widgets per hour. We may say that the first machine is 10% less productive than the second, even though each machine puts out 100% of its capacity. However, if the factory had only one of these machines, its management would view a loss of productivity only relative to that machine's capacity. In other words, if that machine is shut down one hour during a ten-hour production run, there would be a 10% production loss. Obviously, the production loss is measured against the normal output (or capacity) of the available equipment and not against some other equipment that does not exist in that plant.

This understanding is of utmost importance for assessing construction-delay claims. A company usually estimates the labor required for a project by taking account of its own capacity (that is, the experienced productivity of its own workforce), the expected productivity of unknown crews that must be added, and any unusual conditions that may be encountered during construction that may influence the normal or expected productivity. Therefore, it is not valid to use national labor-unit tables or the theoretical ability of another company to try and prove or disprove productivity claims of the company involved in the construction-delay claim. To be validly critical, one must look at the track record of the involved company, and at its estimate: is the estimate error free, and are all conditions that should have been anticipated properly provided for?

By error free, I mean that all materials that had to be purchased for the project have labor allowed for them, that incidental labor that was required in addition to installation labor is properly allowed, and that all extensions and additions are correct. Usually, as a first step, a close comparison of estimated and purchased materials is essential.

Once the error-free condition of the estimate is established or, alternatively, adjustments are made for any errors found, the second step is to establish that the labor units used by the company in its estimate are based on the company's experience, and that the company made all proper adjustments to the estimated installation labor to allow for job conditions that could reasonably be anticipated to differ from the norm.

Last, but not least, it must be established that proper adjustments are made for errors or unusual occurrences during construction that are unrelated to the claim. If this adjustment is not made, the credibility of the claimed production loss is in serious question.

Before we explore the various causes for production losses, I want to offer a cautionary note on labor-unit tables. Labor-unit tables are made up of averages — often averages of averages. For example, let's assume that the labor unit offered by the table for a certain size and weight of pipe for installations up to twelve feet above the floor is four hours per hundred feet of pipe (4.0 hr/C') and for installations from twelve to twenty-five feet above the floor is five hours per hundred feet of pipe (5.0 hr/C'). It is obvious that pipe installed at a twelve-foot ceiling level takes more labor than does pipe installed at an eight-foot ceiling level, which, in turn, takes more labor than that required to install the pipe along walls at a lower level. Similarly, installing the pipe at a fourteen-foot level takes less labor than installing it at a twenty-five-foot level, and so on. Obviously, most "average" labor units consist of a number of averages (the average of averages) and are seldom universally applicable.

Since estimators are notoriously short on estimating time, they often use averages indiscriminately to save as much time as possible. The estimator may not always try to separate the various installation areas to determine the amount of pipe installed at five-foot, eight-foot, or twelve-foot levels, and so on. In fact, the estimator may not even separate his take-off for installations below and above twelve-foot levels; he may simply note that approximately 20% of the total take-off is installed between twelve- and twenty-five-foot levels. This method enables him to use a new average labor unit of 4.2 hr/C' (80% @ 4.0 hr/C' plus 20% @ 5.0 hr/C') for the total pipe take-off. For estimating purposes, this practice is acceptable, providing that the averages used represent the true installation conditions, which is often not the case.

The labor-unit tables may also reflect a mixture of journeymen and apprentices, or they may represent the time for journeymen only. In the former case, training time has to be factored into the labor units. The question is, how much training time is factored in? Some localities allow one apprentice to one journeyman and other localities do not permit more than one apprentice to three or four journeymen. Labor units that are based on journeymen only can be 10% to 20% less than labor units based on a mixture of journeymen and apprentices. Even labor-unit tables based only on journeymen reflect the average of slow and fast workers, and the ratio between slow and fast workers may change from time to time and from one locale to another. It is obvious, therefore, that all such labor-unit averages must be applied with extreme caution and that a careful analysis may be in order.

Now we can look at various causes for production losses and whether or not they may be related to the delay claim. The main thing to remember is that, regardless of whether or not the average labor units from the labor-unit tables used for the estimate are correct, delay causes will always increase the estimated labor.

CAUSES FOR PRODUCTION LOSSES

Most disruptions to the work and the construction schedule cause delays, slowdowns, and production losses. Some production losses should be foreseen when preparing the tender; other production losses can be avoided or largely mitigated; and a few production losses will result in delay claims. However, one must first be aware of their existence. The following is a partial list of causes to watch for:

- **Poor designs:** Whether crews run into stumbling blocks requiring clarifications, or whether designers wish to improve their designs, the work is slowed down and/or disrupted.
- **Lengthy responses to clarification requests:** Poor designs generally lead to clarification requests and/or change requests. If the response takes too long or if there are too many of these requests, the project can be seriously delayed, along with a production loss.
- **Added or deleted work:** When the owner has a change of mind regarding the scope of work, the contractor gets a change notice. Changes disrupt the work in progress while the contractor prepares estimates, waits for clarifications and approvals, orders new materials, and reinstructs the workforce.
- **Poorly ordered materials:** Sometimes not enough attention is paid to ordering or specifying materials for ease of installation. This oversight can cause serious production losses.
- **Late deliveries of materials:** Whoever orders the materials must also take responsibility for specifying timely and accurate deliveries. Untimely and inaccurate deliveries can seriously disrupt and delay the work.
- **Original work schedule too short:** This shortage usually causes one delay after another because all of the task periods are too short; eventually, the work piles up and must be accelerated.
- **Work areas being put on hold:** Work areas may be put on hold for many reasons, generally with associated delays and production losses.

- **Work-sequence changes:** Sometimes, work areas or equipment and materials are not available when scheduled, or the owner may decide on a different sequence from the one specified to turn over the facility. Most work-sequence changes cause production losses. Turn-over priorities can impact the work at the best of times but more so if changes come after the planning stage.
- **Untimely schedule extensions:** Owners are often reluctant to allow schedule extensions, and when schedule extensions become inevitable, they are usually too late for proper planning and may only provide an opportunity for workers to slow down; watch out that work does not expand to fill the new time available (see Parkinson's Law).
- **Mutually adverse effects between trades:** Trades that experience production losses can easily affect others with interdependent activities. This effect is probably one of the most overlooked production-loss causes.
- **Adverse weather conditions:** Each schedule may have some of its work affected by adverse weather conditions, but when disruptions and resulting delays push summer work into winter, the production loss and further delays can be severe.
- **Work accelerations:** Accelerations are usually the result of trying to maintain a slipping or slipped schedule. Accelerations can cause severe impacts because of poorer planning abilities, overmanning, inadequate tooling and other supplies, lowering morale, confusion, and so on.
- **Overmanning:** When contractors must accelerate the work to offset disruptive delays, they often overman the project to avoid overtime. Overmanning not only has its own disruptive effect, but it can also slow down the work so as to reduce the desired acceleration.
- **Overtime work:** Working extended overtime, as numerous studies have shown, can be fatiguing and morale lowering, with resulting productivity losses.
- **Lowered morale:** Repeated disruptions to their work severely lowers the workers' morale, with a resulting slowdown and productivity loss.
- **Workforce supplied from various union jurisdictions:** This problem usually occurs on major projects when local unions are unable to supply the entire workforce and must request workers from other union jurisdictions. It usually causes production losses.
- **Unsafe working conditions:** Unsafe working conditions and

cluttered worksites can severely affect production.

- **Labor disputes:** These disputes usually cause slowdowns and/or work stoppages, with resulting production losses.
- **Location of storage, accommodation, and eating facilities:** If facilities are located too far from the work areas, they can have serious disruptive effects.
- **Contractors' shortcomings:** Shortcomings include unfamiliarity with the contracted work, estimating mistakes, planning mistakes, material-type mistakes, installation mistakes, inadequate tools and construction equipment, or inadequate supervision and project management. The cumulative effect of these shortcomings by all contractors on site can be substantial, resulting in serious production losses. Please note that contractors preparing delay claims must carefully assess these shortcomings when making adjustments to their modified total cost, if this method is used for quantifying the damages.

These are just a few examples of production-loss causes. Some of these causes can and should be anticipated before construction, and others will come later as a surprise. Those that can be anticipated should be properly provided for in the estimates, and those that come as a surprise and result in claims should be properly settled when they occur, during the construction period.

UNIT-PRICE CONTRACTS

As mentioned in Chapter 2, unit-price contracts or task-per-area assessments can be useful in determining exact production losses and in isolating their causes and damages.

In a unit-price contract, the number of installed units is usually approved by the owner on a daily or weekly basis for billing purposes. A spreadsheet can be designed for the labor portion of the units, a) to keep track of the number of units installed in each measured period, and b) to compare the actual labor spent to install these units in the measured period with the estimated labor. One hopes that if there is a variance it will be on the positive side. However, when the spent labor exceeds the estimated labor, a speedy analysis of what went wrong can be made. Following are some of the main causes to look for:

1. The labor units used in the estimate may simply be too low for the tasks, job conditions, and crews involved.
2. Because the labor units are averages, the tasks completed may be on the wrong side of the averages.
3. There may be too much stand-by time or study time between performing the tasks.

4. There may be too many disruptions while the tasks are being performed.
5. There may be a combination of the above.

A good supervisor can usually determine quickly which of these causes is involved, and then initiate the appropriate corrective measures. The point is, because a unit-price contract usually shows up deficiencies with the performance quickly, why not turn every project into a unit-price contract — even if this is done only internally, within the contractor's own administration, and, perhaps, through the Task-per-Area Assessment method?

The benefits are obvious: first, the contractor can determine quickly how accurate his labor units really are; second, he can also quickly determine if and when any causes for lost time occur; third, he will be in a better position to give timely notices to the party or parties at fault for the lost time; fourth, he will be in a better position to determine the exact amount of lost time; and, fifth, he can take corrective and mitigating measures faster. Much is to be gained for relatively little more administrative work.

Another advantage of unit-price contracts is that they usually give a definite percentage (for example 25%) of quantity changes that can be made without requiring adjustments to the unit prices; therefore, the argument of excessive changes to the contract is easier to settle.

EXAMPLES OF PRODUCTION-LOSS CLAIMS

Example 1: *Demobilize, Remobilize, and Retrain Workers*
An owner notifies a contractor to stop work in five days. The contractor has 80 construction workers on site. The site must be cleaned up and secured, materials and tools must be stored, work areas must be left in a safe condition, and arrangements must be made for surveillance of the abandoned site.

After three months, the owner notifies the contractor to continue with construction. The contractor is able to rehire 40 of the 80 laid-off construction workers but must hire 40 new workers. The site must be prepared for work again: materials must be checked and redistributed, tools made ready, and so on. The rehired workers must refamiliarize themselves with the site conditions and the work to be done, and the new workers must be trained.

The contractor submits a claim for lost time because of demobilization at four hours per worker, remobilization at four hours per worker, refamiliarization of the rehired workers at eight hours per worker, training of new workers at 24 hours per worker, plus the various site and company overheads during the dormant period, including the surveillance cost.

The owner, after haggling a bit, agrees to pay most of the claim. Why? Because this is a fairly straightforward situation. The cause and effects of the damages are highly visible and easily quantifiable, and the contractor has acted reasonably under the circumstances.

However, isolated instances such as this example are rare, and when damage causes and their effects are less clear and become harder to prove, owners can be stubborn during settlement negotiations and often reject the claim outright.

Example 2: *Time Lost to Correct Poor Designs*
A tender request called for the installation of owner-supplied materials and equipment at a remote plant. A list of these materials and equipment was issued with the tender documents, and the bidders were asked to allow only for minor additional materials to make the installation complete.

The owner provided the camp for the construction workers free of charge for the number of worker-camp-days specified by the tenderer, and used $60 per day for these specified days to supplement each tender price for comparison purposes. (The charge to the contractor for days used in excess of his specified days was also $60 per worker-camp-day.)

Three tenders were submitted. Contractor A was chosen to do the work. All tenderers were aware that the local unions were unable to supply the entire workforce required and would have to request sister unions to help out. Contractor A applied national, standard labor units to the materials and equipment, and then added 10% for the unknown-crew factor. He also knew during the tender period that the designs were poor but decided not to factor this in, because he felt that any impact caused by having to correct poor designs should become an added charge to estimates for required change orders (which the owner later rejected after each quote submission).

During construction, the work was constantly disrupted by the necessity of clarification requests and necessary change orders to correct poor designs. The contractor gave early notice to the owner of a pending claim for lost time due to these disruptions, which was disregarded by the owner. At the end of construction, the contractor submitted his claim, which was eventually also rejected by the owner.

Contractor A's labor estimate, including the 10% for the unknown-crew factor, had been 100,000 worker-hours, and his actual labor ended up at 114,900 worker-hours. In other words, he had only achieved an 87% productivity level, which he blamed entirely on the disruptions caused by the poor designs. The contractor and the owner had agreed, at the outset, on a charge of $65 per worker-hour for extra work. This charge included the direct labor cost, the owner's camp charge, all other labor burdens, the contractor's overheads, and profit. The contractor's claim was based on the 14,900 worker-hour overrun at $65 per worker-hour, for a total of $968,500.

Before rejecting the claim, the owner contacted the other tenderers and found out that both had used the same national, standard labor units, but contractor B had used 20% and contractor C had used 25% as the labor adders for both the unknown crew factor *and* the poor designs, which, they had expected, would cause substantial disruptions to the work with or without corrections.[*]

In his explanation for the rejection of contractor A's claim, the owner reminded the contractor that the issue of poor designs had been raised at a pretender meeting, and all potential tenderers had been cautioned that this was a fast-track project with partially incomplete designs; therefore, contractor A should have made a proper allowance for this factor in his original tender.

Contractor A rejected this explanation, claiming that an accurate assessment of disruption impacts can only be made at the time of occurrence; this assessment, he said, he had done at the time when he submitted his quotes for the corrective, extra work, but the owner had rejected and excluded his impact charges.

This case is not unusual. The facts are not in dispute, and the arguments, pro and con, have some logical validity. How is a contractor to know at the tender stage to what extent, if at all, an owner intends to correct his poor designs or disallow the contractor's impact charges? On the other hand, poor designs, even if left uncorrected, will certainly cause production losses, and the owner's expectation that an allowance for these losses should be made by the contractor in his original estimate is justified. In the final analysis, this sort of claim may either end up to be a hat-in-hand claim or may only be resolvable on legal grounds.

Example 3: *Subsurface Obstructions versus Adverse Weather*

A contractor contracted the installation of underground duct lines. He had scheduled the work to be done with equal manpower loading during the months of August, September, and October, using 24 workdays in each month, and submitted this schedule with his tender. The owner accepted the tender without comment.

When the contractor started the work, he immediately ran into some

---

[*] Note that both percentage adders had resulted in estimates below contractor A's actual labor hours, although contractor C's estimate came close — he would have experienced a 98.9% productivity level with his estimate.

old foundations that were buried beneath the surface. He gave the owner notice, since the contract provided relief for such circumstances. These buried foundations were encountered throughout the construction period, and the contractor finished the work just before Christmas, 39 workdays past the end of October, his original schedule.

The contractor promptly submitted his claim for the additional 39 workdays, and the owner promptly rejected it. The owner explained that, although he admits having some responsibility for the delay because of the obstructions encountered by the contractor, he has had a chance to check the contractor's projected manpower loading with existing industry standards and has found them to be below these standards. Furthermore, the owner pointed out, the contractor had allotted equal portions of the work in his schedule with equal manpower for each of the scheduled construction months, and his estimate had, therefore, made no allowance for the extremely cold weather that was encountered during October.

Since the contractor had no comeback on the owner for productivity losses due to cold weather conditions in October, he went back to the owner a second time and offered to make a reasonable adjustment in the claim for this factor. The contractor pointed out, however, that even during the normal-weather months of August and September he had experienced only 80% productivity, which was clearly due to the subsurface conditions he had encountered, and this loss had pushed some of the work planned for August and September into the cold-weather months of October, November, and December.

The owner was not entirely convinced by this argument, and insisted that the contractor could have experienced part of his alleged productivity loss as a result of low labor units allowed in his estimate; but to show his willingness to be fair, the owner offered the contractor one-third of the claim amount in settlement on a no-prejudice basis.

The contractor was desperate, because he could not afford such a heavy loss; but he also had some doubts about whether or not he would be able to convince an arbitrator or a judge of more than half of his claim. He decided to seek the advice of a claims expert.

The claims expert looked at the tender, the contract documents, the record of actual labor expended during the various periods, and the statistical weather records during the months of October, November, and December for the year of construction compared with other years. He found that the month of October was indeed much colder than normal, but the months of November and December were equally as cold as in normal years. His statistical productivity information on the various temperatures, winds, and humidities encountered confirmed that, at best, the contractor could have expected a 66.7% production during these months.

And since the contractor had experienced no normal period during the entire construction duration as far as productivity was concerned, the expert investigated several similar projects that the contractor had successfully completed, using the same crew and labor units, and found that indeed, even though the contractor used labor units below the industry standard, his crew and his progressive construction methods allowed him to do this work successfully within the estimate.

The claims expert wrote his report with the following explanation: Had the contractor not encountered any subsurface obstructions, he would obviously have only been retarded by the cold weather conditions in October, which would have lost him 12 workdays; that is, the 24 workdays he had allowed divided by the 0.667 productivity factor for these cold weather conditions encountered equal 36 workdays required to do the work. And these 12 workdays lost are all that the contractor is responsible for under the contract.

The expert decided to include the following explanation and a table (see Table A-1) for the benefit of the owner. The constant disruptions encountered because of various subsurface obstructions had lowered the

	Workdays	Time Loss
Contractor's Original Schedule		
(based on normal weather)		
August	24.0	0.0
September	24.0	0.0
October	24.0	0.0
Total	72.0	0.0
Schedule Adjusted for Weather Only		
(contractor's responsibility)		
August	24.0	0.0
September	24.0	0.0
October (@66.7% Productivity)	24.0	8.0
November (@66.7% Productivity)	12.0	4.0
Total	84.0	12.0
Actual Performance		
(impacts due to obstructions & weather)		
August (@80% Productivity)	24.0	4.8
September (@80% Productivity)	24.0	4.8
October (@80% times 66.7% Productivity)	24.0	11.2
November (@80% times 66.7% Productivity)	24.0	11.2
December (@80% times 66.7% Productivity)	15.0	7.0
Total	111.0	39.0

Table A-1 Time-loss breakdown

contractor's productivity to 80% during August and September. Had he not experienced cold weather in October, the time lost during August, September, and October would have amounted to 18 workdays; that is, 72 scheduled workdays divided by the 0.8 productivity factor equals 90 workdays, 18 workdays more. These 18 workdays would have been pushed into November, and the cold weather conditions in this month would have lost another 9 workdays; that is, 18 workdays divided by the 0.667 cold-weather productivity factor equals 27 workdays. Thus, the expert proved that these 27 workdays lost is what the owner is responsible for under the contract.

The contractor resubmitted his claim on this basis, asking for reimbursement of only 27 workdays production loss. He also offered the owner the services of his expert to provide the appropriate explanations. Following another two meetings with the contractor, the owner accepted the revised claim.

Example 4: *Delays, Accelerations, and Adverse Weather*

A contractor signed a contract to build an industrial project in 10 months. The owner specified a number of major materials and equipment to be supplied by him at certain times during the construction schedule. The contractor has had extensive experience with the type of work involved, and his average crew size to perform the work was estimated at ten workers. The original contract price breakdown is shown in Table A-2.

From the outset, the owner failed to make the work areas available in time to meet the construction schedule, and some of his material and equipment deliveries were also delayed. Most schedules depend on areas being available in time to perform the work, materials arriving on time to meet work sequences, sufficient manpower to attain milestone and completion dates, and minimal disruptions to the work. The contractor notified the owner each time another delay occurred, but conditions did

Original Contract Breakdown	
Construction Materials by Contractor	$882,000.00
Labor (10 workers @ 160 hr/mo x 10 mo @ $48/hr)	768,000.00
Project Overheads (for 10 mo)	120,000.00
General Overhead (for 10 mo)	120,000.00
Profit (based on 10 mo)	60,000.00
Contract Price	$1,950,000.00

Table A-2

not improve: the disrupted schedule caused a number of production losses due to rescheduling work sequences, demobilizing and remobilizing crews in several areas, and inevitable waiting and stand-by time, with resulting loss of worker morale. He tried to mitigate the impact on his production by reducing the crew several times. Nevertheless, his revised construction schedule came up with a two-month completion delay and an average crew of nine workers (i.e., an 8% budget overrun). He notified the owner of this delay and provided an estimate of the extra cost as per Table A-3. The owner objected to the revised completion date but issued no orders to accelerate the work.

After ten months of construction, it became evident that an early winter with adverse weather would affect the work that had been pushed into the last two months, 77.78% of which was outside, exposed to the weather. The contractor knew that the best productivity under these circumstances would drop to 80% of normal for the remaining exterior work. And because he did not want to propose another completion delay to the owner, he decided to hire two more workers for the remaining 1.75 months.[1] He also notified the owner of this development and submitted a revised delay claim, with restated production losses and damages as per Table A-4.

The owner, who had stubbornly refused to grant the contractor an extension to the completion date, now realized that the two-month delay was inevitable. He quickly called a meeting to discuss alternatives with the contractor. The contractor told the owner that adding more manpower was impracticable, and the owner requested the contractor to work his crews overtime to accelerate the work.

The contractor then worked out an agreeable overtime schedule with his supervisor and his crew. They agreed to work 20 hours of overtime each week starting in the second week of the eleventh month. As a result of this overtime, the project was completed in five weeks, saving the owner two weeks of lost operating time.

However, the contractor had lost a further 6.67% overall productivity during the five-week overtime period,[2] which he included in his claim, although he saved some project overheads, general overhead, and profit charges on lost revenue. He resubmitted his delay

---

[1]
Work left in remaining two months	=	2,880 hr
Work to be done outside, in normal weather	=	2,240 hr
Outside work in winter (@ 80% efficiency)	=	2,800 hr
Overrun (= 2 workers @ 1.75 mo @ 160 hr/mo)	=	560 hr

[2]
11 workers working 20 hr overtime for 5 weeks	=	1,100 hr
Time saved = 11 workers @ 40 hr for 2 weeks	=	880 hr
Time lost due to overtime	=	220 hr

Delayed Contract Breakdown

Construction Materials by Contractor	$882,000.00
Labor (9 workers @ 160 hr/mo x 12 mo @ $48/hr)	829,440.00
Project Overheads (for 12 mo)	144,000.00
General Overhead (for 12 mo)	144,000.00
Profit (based on 12 mo)	72,000.00
Total	$2,071,440.00
Less Original Contract Price	1,950,000.00
Delay Claim	$121,440.00

Table A-3 Delay production loss and damages

Delayed Contract Breakdown

Construction Materials by Contractor	$882,000.00
Labor (9 workers @ 160 hr/mo x 12 mo @ $48/hr)	829,440.00
Labor (2 workers @ 160 hr/mo x 1.75 mo @ $48/hr)	26,880.00
Project Overheads (for 12 mo)	144,000.00
General Overhead (for 12 mo)	144,000.00
Profit (based on 12 mo)	72,000.00
Total	$2,098,320.00
Less Original Contract Price	1,950,000.00
Revised Delay Claim	$148,320.00

Table A-4 Delay plus weather production losses and damages

Delayed Contract Breakdown

Construction Materials by Contractor	$882,000.00
Labor (9 workers @ 160 hr/mo x 10.25 mo @ $48/hr)	708,480.00
Labor (11 workers @ 240 hr/mo x 1.25 mo @ $48/hr)	158,400.00
Premium (11 workers @ 80 hr/mo x 1.25 mo @ $24/hr)	26,400.00
Project Overheads (for 11.5 mo)	138,000.00
General Overhead (for 11.5 mo)	138,000.00
Profit (based on 11.5 mo)	69,000.00
Total	$2,120,280.00
Less Original Contract Price	1,950,000.00
2nd Revision Delay Claim	$170,280.00

Table A-5 Delay, weather, plus acceleration production losses and damages

claim as per Table A-5, and the owner, after carefully checking the breakdown, accepted the claim.

It is interesting to note that there were few changes to the work, and the contractor completed them on a cost-plus basis with his maintenance crew. Had there been changes to the contract, the contractor would have assessed them individually for the impact caused to the original contract work and would have submitted a claim for the impact in a similar fashion to those above.

POSTSCRIPT

I chose Examples 3 and 4 to illustrate how a number of production-loss causes can be separated to show the resulting individual damages. This is a desirable method for convincing the party asked to pay for delay claims, but it is not always practiced by contractors. Knowledge of production-loss causes and damage quantifications will make this practice easier to follow.

Glossary
of Construction Terms
and
Related Concepts

GENERAL COMMENTS

The following glossary lists definitions that the author had in mind when writing the book. The terms and phrases used may or may not comply with standard dictionary terms and phrases, because they are intended to apply mostly to construction and construction delay claims. Terms and phrases shown in *italics* are defined in this glossary.

alternative dispute resolution — an alternative method to trial by court to resolve the dispute.

arbitration — one *alternative dispute resolution* method, usually done in less time and at less cost than a *trial* and usually resulting in a binding decision on the parties.

borrowed capital — the capital a company must borrow for its operations or its investments.

budget excess — the amount by which a budget such as the labor estimate is exceeded by actual expenditures.

claim cause — any cause resulting in a *claim submission* for damages.

claim submission — payment demand for *claim cause(s)* and *effect(s)*, including supporting evidence, issued by a *claimant* to another party, the *respondent*, who is allegedly responsible for the cause(s), and the damages incurred.

claimant — the person or company issuing a *claim submission*.

claims expert — one who specializes in technical opinions on, and the preparation of, *construction delay claims*.

claims lawyer — a lawyer who specializes in legal opinions on, and litigation of, *construction delay claims*.

conciliation — a nonbinding method of *alternative dispute resolution* usually involving *mediation*.

constraints assessment — a time and production loss evaluation, usually completed by a field supervisor, for any work disruption, slowdown, or delay assessed to a blamed party.

construction delay claim — a damage claim for schedule delays, *work* slowdowns and/or disruptions, either submitted directly to a *respondent* or filed with the Clerk of the Court as a lawsuit.

construction schedule — a planned time frame for completing various construction tasks and the project.

contingency — a potential known or unknown risk.

contingency allowance — an amount allowed for potential risks, often included with a *profit* charge.

contract — an agreement to perform specified construction *work* upon specified terms, and usually in a given time frame.

contract conditions — the conditions under which a *contract* must be completed.

contractor — the person or company who carries out the conditions of a *contract*, either for an *owner* or another contractor.

delay overheads — *overheads* attributable to construction delays.

deposition — evidence given under oath by a construction dispute witness, usually at *discovery*.

differential measurement — method of measuring *productivity* by comparing "normal" with "impacted" production.

discovery — a legal means for each party to the litigation of finding out the strengths and weaknesses of the other party by examining their witness(es) under oath.

dispute — or construction dispute — a disagreement over the cause and/or effect of an experienced or potential damage.

documentation — documents forming part of the tender and contract, the written communications between the parties, time sheets, diaries, and other records, relative to a project.

economy of scale — the reduced cost resulting from a higher quantity.

effect — the result, which is usually the impact on production and/or time and the associated costs, of some cause of disruption or slowdown of the work.

Eichleay formula — a formula for allocating additional *general overhead* to a delayed project; it is based on the ruling of the U.S. Armed Services Board of Contract Appeals, decision 5183, December 27, 1960, in an appeal by the Eichleay Corporation from a ruling of a contracting officer, and is now widely used in the USA.

employee benefits — benefits such as bonuses, holiday and vacation pay, contributions for pensions and health insurances, and so forth, accruing to the employee in addition to his basic wages or salary, and forming part of the company's overall *labor burden*.

entitlement — the condition(s) of a *contract* giving a party a legal right to a claim for damages.

estimate — an approximate evaluation of the construction costs of a *tender*.

executive summary — short summary of the causes, *effects*, and damage *quantifications* of a claim.

extended guarantee — a warranty beyond the normal guarantee period, usually required by delayed project completion.

field supervisor — a manager of field operations, usually in direct contact with the workforce.

general contractor — a contractor contracting on behalf of all, or a significant number of, trades.

general overhead — a contractor's overhead required for marketing, usually allocated to the company's various projects by a formula.

guarantee — a promise by a supplier or a contractor that the product or installation will function as specified for a certain period.

holdbacks — the money withheld by an owner from a contractor's invoices, sometimes to guarantee performance but usually to satisfy lien legislation.

instructions to bidders — the *owner's* or owner's agent's instructions to the tenderers of the requirements of a tender and potential contract.

interest charges — finance charges encountered by the delay of contract payments and holdbacks for work done and/or materials supplied.

interest versus return on investment (ROI) — interest is a *special overhead*, at the reduction of ROI if *borrowed capital* is required for *working capital*; otherwise, it is a *general overhead*, and the full ROI applies.

intermediary — a person acting as a mediator to assist in the negotiations between the parties to a claim.

job factors — various factors for labor-cost adjustments that must be added for adverse weather and other conditions peculiar to each job.

job progress — the progress of the contracted work, usually expressed as a percentage of the total.

job-related overhead — same as *project overhead.*

labor burden — the total labor-cost additional to the basic wages or salaries, usually including costs such as insurances in addition to *employee benefits.*

labor efficiency — a production with a minimum of wasted effort; often confused with *labor productivity.*

labor escalation — the periodic inflationary increase to the labor cost.

labor productivity — the amount of production achieved relative to the labor-cost estimate: for example, if the budget is exceeded by 10%, the productivity level is at 90.9%; not to be confused with *labor efficiency:* a crew can be at a 100% productivity level, yet inefficient in its work habits.

learning curve — a worker's gradual increase of productivity during the job and process familiarization stage.

manpower loading — the number of workers required to complete the work in the construction scheduled at any given time.

material escalation — the periodic inflationary increase to the material cost; usually affects only materials not yet ordered.

measured mile — a method of measuring labor productivity levels by comparing a normal construction period with an abnormal one for similar tasks.

mediation — one alternative disputes resolution method of reconciling the differences between disputants,

usually nonbinding on the parties.

mitigation — the actions taken by an affected party to lessen the damages caused by work disruptions and delays.

modified total-cost method — a method of quantifying the cost overrun due to delays by making adjustments to the total cost for the affected party's own shortcomings.

overbilling — invoicing for materials and/or labor not yet delivered.

overhead — all costs in addition to a company's labor, material, and subcontract costs; overhead in construction normally consists of *general overhead* and *project overheads.*

overhead allocation — a method of allocating *general overhead* to a project, which is especially useful for delayed projects.

overhead efficiency — a company's level of standing compared with an industry average.

overhead escalation — increased overhead costs due to inflation and project completion delays.

owner — the person or group of persons who own, will own, or operate the construction facility.

ownership capital — the equity capital invested by a company's shareholders.

Parkinson's Law — an observation that work expands so as to fill the time available for its completion, first published by C. Northcote Parkinson in Great Britain, 1958.

prime consultant — an owner's consultant who heads up a group of the owner's consultants.

production-loss allocation — the assignment of a production loss to a party alleged to be responsible for it.

productivity — the level of actual labor cost compared with the budgeted labor cost; if the former exceeds the latter by 10%, the productivity achieved is 90.9%.

productivity loss — the difference between the lower productivity achieved and 100%.

profit — a company's charge for the use of *ownership capital*, usually including a *contingency allowance* for unknown risks.

project manager — a person who administers the contract conditions and construction schedule for an owner or a contractor.

project overhead — all overheads of a company chargeable to a particular project, which would not exist without the project.

project-related overhead — same as *project overhead*.

quantification — measurements and calculations done to establish the damages caused by work slowdowns, disruptions, and delays.

respondent — the recipient of a *construction delay claim*.

risk allowance — a provision for potential risks, either a specific dollar amount if the risk is known or an increased profit allowance if the risk is unknown and/or inexact.

risk assessment — an estimator's estimate of the potential risk.

schedule — see *construction schedule*.

schedule delay — a delay of a scheduled task and/or the completion of the project.

schedule disruption — a disruption of the *work*, causing a *schedule delay*.

schedule sequence — the most logical, effective, and economical sequence of performing the various tasks of the scheduled *work*.

settlement — the final outcome of a construction dispute.

snapshot technique — a method of determining the delay, if any, at the end of each task of the *work*, and the cause(s) and effect(s) of the delay.

special overhead — same as *project overhead*.

subcontractor — a contractor who has a contract for work with another contractor, usually a *general contractor*.

subsubcontractor — a contractor who has a contract for work with a *subcontractor*.

task-per-area assessment — a method of assessing the periodic completion percentages of various tasks for the *work* and the entire project.

tender — a price proposal submitted to the owner, owner's representative, or another contractor for specific *work* required by them.

tender qualifications — conditions stipulated in a *tender*, sometimes altering the project specifications, that apply to make the tender valid.

Thormann formula — a method of allocating a company's *general overhead* to its projects in proportion to its *special overheads* (equitable for delayed projects).

total-cost method — taking the total cost of a project at completion of the *work* less the estimated cost and assigning the difference to the claim as the damages of the *claim causes*.

trial — a lawsuit conducted in court.

triangle of performance — the notion that performance is lacking if a worker cannot do the job, does not know how to do it, or does not care.

work — 1. with respect to contracts, the materials, labor, and plant required to build the specified construction project; 2. with respect to workers, the result or expected result of their efforts.

workflow — see *schedule sequence*.

working capital — assets used in the everyday financial operations of a company.

work-in-progress — work undertaken but not yet completed.

Bibliography
of
Related Subjects

PLANNING AND SCHEDULING

Kerzner, Harold. *Project Management: A Systems Approach to Planning, Scheduling, and Controlling*, Second Edition. New York: Van Nostrand Reinhold Company (1984). This book has some excellent chapters on planning, on the program evaluation and review technique (PERT), on the critical path method (CPM), and on the bar chart (GANTT), including the required project graphics. A good understanding of this process is extremely important to evaluating delay situations.

GENERAL PRODUCTION-LOSSES

Blough, Roger M. *More Construction for the Money: Summary Report of the Construction Industry Cost Effectiveness Project*. New York: The Business Round Table (January, 1983). This book presents a study of various shortcomings of management as well as organized labor with respect to productivity losses. It also offers some remedies for the reader's consideration. The value of the book lies in the recognition of production-loss causes (with resulting delays) that may be inherent with a project and, therefore, not claimable.

National Electrical Contractors Association (NECA). *Manual of Labor Units*. Washington, D.C. (1995). This manual has very useful appendices that include a number of different production-loss studies for various work conditions, for example, the effect of multi-story buildings on productivity, the effect of temperature on productivity, and the effect of overtime on productivity.

Parkinson, C. Northcote. *Parkinson's Law or The Pursuit of Progress*. London, Great Britain, John Murray (1958). I included this reference because I believe Parkinson's Law explains one of the more insidious causes of production losses. I believe it is the explanation for the drop in productivity that occurs when seemingly non-vital delays (of information, work areas, materials, tools, etc.) cause employees to keep busy. This phenomenon deserves much more attention and study than it has hitherto received from construction personnel.

OUTPUTS FOR VARIOUS HOURS OF WORK

Bureau of Labor Statistics. *Hours of Work and Output.* Washington, D.C.: United States Department of Labor (1947). This is a very interesting study conducted by the U.S. government during WW II on the output effects that were experienced for changes in the standard workweek of 40 hours. The value of this study has been questioned in the 1990s, because various changes in present working conditions (longer coffee breaks, technological advantages, etc.) have also brought changes to the fatigue factors.

CHANGE-ORDER PRODUCTION-LOSS

Gander, Leon P. *The Building Construction Process Explained in Plain English.* Burnaby, B.C.: Construction Reality Publishing Inc. (1999). Leon P. Gander has been in the construction industry since he graduated from the University of New South Wales, Australia in 1968. He is a Professional Engineer with an electrical consulting practice in Vancouver, British Columbia. His book provides not only an excellent description of various processes of construction proposals and construction, but also of the effects of errors and omissions in designs and the inevitable change orders to correct them — with resulting delays of construction and losses of productivity.

Construction Industry Institute. *The Impact of Changes on Construction Cost and Schedule.* Publication 6-10. Austin: Bureau of Engineering Research, The University of Texas at Austin (April, 1990). This booklet provides many valuable details of construction changes and how they affect the construction schedule and cost — especially direct and consequential impacts of changes. It lists a number of recommendations and ideas to consider as well as a page of references on related subjects.

THE EFFECT OF TEMPERATURES ON PRODUCTIVITY

National Electrical Contractors Association (NECA). *The Effect of Temperature on Productivity: Test Report.* Washington, D.C. (1974). This report provides us with productivity levels at various relative humidity levels and effective temperatures, that is, wind-chill-factor adjusted temperatures.

EXAMINATIONS FOR DISCOVERY

Shapiro, Stuart. *How to Survive a Deposition*, Mark Siegel, Ph.D., editor. New York: John Wiley & Sons (1994). This is one of the best books on the complete discovery process. Much valuable insight is given to the motivations behind some of the questions asked by opposing counsel. It also provides extensive coaching for witnesses.

CONTRACTS, BREACHES OF CONTRACT, AND DAMAGE ENTITLEMENTS

Goldsmith, Immanuel (Q.C., LL.B., London) and Heintzman, Thomas G. *Goldsmith on Canadian Building Contracts, Fourth Edition*. Toronto: Carswell (1988). This publication first came out in a hard-cover edition in 1968, with references to just under 1000 cases. The present edition is a loose-leaf binder. The case references have expanded to over 2300, and the binder is periodically updated. It is an excellent publication on various contracts, contract breaches, damage entitlements*, and applicable case references.

Revay, Stephen G. *Construction Law Letter, Special Report 4: Construction Claims: Causes and Options*. Toronto: Build/Law Publications Inc. (ca. 1993). This book is similar to *Construction Delay Claims* in its description of the elements affecting claims, but it focuses more on entitlements. *Construction Delay Claims* gives more illustrations of the various quantification fallacies. Revay's book presents valuable additional reading material.

AVERAGE LABOR UNITS

Winslow, Taylor F. *Construction Industry Production Manual: Accurate Labor Tables for All Construction*. Solana Beach, CA: Craftsman Book Company (1972). This manual provides reasonable labor-unit averages for the sixteen disciplines that make up construction work. Although averages can be deceiving and must be frequently adjusted for individual conditions encountered, this manual can serve as a guide to realistic labor units under normal circumstances.

---

* Our moral indignations and sense of right and wrong are no substitute for the reality of our legal entitlements; therefore, we should know what these are before launching expensive lawsuits.

DISPUTE PREVENTION AND RESOLUTION

The following publications by the Construction Industry Institute's Dispute Prevention and Resolution Research Team provide interesting studies and are recommended reading as supplements to this book.

Disputes Potential Index. Special Publication 23-3. (February 1995)

Dispute Prevention and Resolution Techniques in the Construction Industry. Research Summary 23-1. (October 1995)

Prevention and Resolution of Disputes Using Disputes Review Boards. Implementation Resource 23-2. (June 1996)

Index

of Construction Terms and Related Phrases

About the Author

Arthur O.R. Thormann was born 1934 in Berlin, Germany, and emigrated to Canada in 1951.

He has been involved with construction projects for over half a century and as a consultant on delay claims for a good part of that time. He was also active in labor negotiations, as a trustee on employee benefit funds, and in the teaching of estimating courses. He believes that Canadians are among the freest thinkers of the world and a great people, and Arthur decided to join them as a fellow citizen.

His home base is in Edmonton, Alberta, Canada.

www.ingramcontent.com/pod-product-compliance
Lightning Source LLC
Chambersburg PA
CBHW052109230326

41599CB00054B/5217